FOOD ADDITIVES AND
HYPERACTIVE CHILDREN

FOOD ADDITIVES AND HYPERACTIVE CHILDREN

C. Keith Conners
Children's Hospital National Medical Center
Washington, D.C.

PLENUM PRESS • NEW YORK AND LONDON

Library of Congress Cataloging in Publication Data

Conners, C   Keith.
  Food additives and hyperactive children.

  Includes index.
  1. Hyperactive children. 2. Food additives—Toxicology. I. Title.
RJ506.H9C65   618.92'85890654   80-66

ISBN-13: 978-1-4684-3688-4   e-ISBN-13: 978-1-4684-3686-0
DOI: 10.1007/978-1-4684-3686-0

© 1980 Plenum Press, New York
A Division of Plenum Publishing Corporation
227 West 17th Street, New York, N.Y. 10011

Softcover reprint of the hardcover 1st edition 1980

To Anthony, Rachel, Becky, and Sarah, and to Karen

PREFACE

The purpose of this book is to present an account of several different studies of the relationship of food additives to child behavior and learning problems. Because the outcome of these studies has deep, personal interest for many parents and their children, I have tried to present the studies in such a way that the logic and evidence of the studies is comprehensible to informed adults interested in weighing the facts for themselves. Unfortunately, the facts do not always follow a straightforward course. Part of my purpose has been to show the complexities lying in the way of the answers to apparently simple questions. I believe it is healthy and important for parents to examine the scientific evidence on issues affecting their daily lives, and to become aware of the process of research surrounding controversial claims regarding new therapies.

New ideas in behavioral science are often difficult to track down and evaluate, and consequently there may be a large gap between therapeutic claims and evidence bearing on those claims. The mother who wonders whether her child should be treated with a special diet is unlikely to have the facts necessary to make a judgment of the costs and benefits. She should however, know some of the major pitfalls in coming to a conclusion for or against such a course.

The professionals—psychologists, pediatricians, teachers—who are asked to render judgment on the food additive controversy, will find themselves immersed in a welter of contradictory studies and claims. Unless they have had specific training in clinical behavioral research, they are unlikely to be aware of the numerous pitfalls awaiting data collection relevant to therapeutic efficacy and side effects. I have frequently been astounded, when presenting some of the studies in this book to audiences of physicians and others working with behaviorally impaired children, at how easily one's own anecdotal experience outweighs substantial evidence collected under relatively controlled conditions. I have presented a carefully documented study having a negative outcome, only to have a doctor stand up later and say, "But I had this case of a child who became a different person after only five days on the diet. . . ." Or a mother who will say, "But my son turns into a little monster if he so much as sniffs a red gumball. . . ." Or contrarily, I have presented a positive outcome study only to have a member of the audience say, "Well, I tried it on five kids and it didn't work."

It is precisely because such anecdotes can be *misleadingly* convincing that consumers must learn the rudiments of experimental logic and must become familiar with some of the threats to the validity of a claim, no matter how charismatic its advocates, or how much the weight of authority is cited in support of the claim. Science is not an end state but a process. No single study is usually sufficient in clinical therapeutics to establish the cause or the cure of a problem. Often the context in which assertions are put forth—hallowed halls of the academe, prestigious medical meetings, public hearings in legislative committees—creates more confidence in the findings than is deserved upon careful inspection of the logic of the argument being set forth.

On the other hand, science tends to be a conservative process, and it is not at all uncommon to have important discoveries ignored until the proper atmosphere and climate of opinion make scientists receptive. The idea that food additives cause behavioral and learning impairments is a genuinely new idea within the confines of the research on hyperkinesis, learning disability, minimal brain dysfunction, and behavior disorders in children. The implications of a *behavioral* as opposed to a carcinogenic impact of additives in foods is startling. Food is something everyone ingests throughout the life span, and if small amounts of additives produce even minor perturbations of development or learning, this finding must be regarded with great seriousness. Only recently have the food industry and the government begun to take seriously the *behavioral* toxicity of chemical ingredients in foods (including certain foods themselves). I believe that the episode of "food additives and hyperkinesis" will be a significant chapter in the history of behavioral science because it has turned our attention to an entirely new class of influences on behavior and learning. This novel hypothesis may turn up quite unexpected results—for different reasons than were stated in the original form of the hypothesis. It is hoped that the studies reported in this book will convince parents to evaluate carefully all claims that a particular diet or treatment can help their child, and that professionals studying children will have a firmer basis on which to make professional, clinical judgments and do further research.

C. Keith Conners

ACKNOWLEDGMENTS

David Boesel of the National Institute of Education first brought my attention to the problem of food additives and hyperkinesis. Initial work on the problem was supported by an NIE contract (NIE–R–74–0007). Further work was supported by a grant from the National Institute of Mental Health Psychopharmacology Research Branch (RO1–MH–28458). My collaborator, Charles Goyette, Ph.D., helped me through all stages of the problem and deserves major credit as the principal coordinator of many facets of the several studies reported here. In Boston, Alan Cohen, M.D. and Janice Hubka, M.A. participated in early phases of the program. Joanna Dwyer, D.Sc. gave valuable advice regarding dietary control and aspects of study design. Benjamin Feingold, M.D., who provided the initial impetus for all these investigations, served as an unpaid consultant in the Boston studies.

In Pittsburgh, James Lees, M.D., Paul Andrulonis, M.D., and Theodore Petti, M.D. served as child psychiatrists for screening and diagnosis. Deborah Southwick, M.S. and Patricia Harper, M.S. served as nutritionists in the challenge studies, and contributed much original work to the dietary aspects of the program. Lynn Curtis, Margaret Klecha, Marianna Flowers, Alan Delamater, Stana Paulaskas, Sheri Hanley, Shelly Wolynn, Barbara Seger, and Marjorie Grieshop all provided research assistance. Richard Ulrich, M.P.H. gave invaluable statistical consultations throughout my stay at Pittsburgh. Elisa Newman and Alice Sherk carried out major parts of the learning and follow-up studies, respectively, during elective time from their medical curriculum. Helen Lang typed much of the manuscript with great patience and contributed in numerous ways to the smooth functioning of the reserach team. Cathie Calvert has helped edit and arrange most of the text and also carried out helpful library research.

William Darby, M.D., Ph.D., of the Nutrition Foundation provided funds to support the packaging of the challenge materials, and Wayne Swett of Nabisco kept the codes and packed the double-blind challenge cookies.

CONTENTS

BACKGROUND OF THE PROBLEM

Childhood Behavior and Learning Problems

The largest single group of children at psychiatric clinics is made up of children with behavior and learning problems. Over the years such children have proved to be an enigma. As many as 60–70% of the clientele at child guidance clinics are referred because of problems in management by parents and teachers. Attempts to correct the children's behavior by the usual means have failed; they seem to disregard punishment or praise, except in the short run. Rather than growing out of the problem, their difficulties in impulse control and learning persist into adulthood. Sometimes easily confused with the child who is merely undisciplined, close inspection reveals fundamental impairments in perceptual, cognitive, and behavioral functions that go beyond simple "brat behavior."

For convenience and clarity, it is usual to include children in the category of behavior problems who have normal intellectual function. That is, unlike the mentally retarded child, there is no overall slowness in cognitive functions such as verbal and spatial reasoning ability. Despite their average or better intelligence, however, these children appear stupid to the unsuspecting teacher or observer because they fail at learning tasks. As they fail repeatedly in school, their motivation to learn is diminished and they are then assumed to be merely lazy. But careful laboratory studies have demonstrated that even under optimum motivation, the child's attention will be limited, especially when the tasks are

1

paced by others (Sykes, Douglas, & Morgenstern, 1973). Intellectual performance is spotty and uneven as though there were selective deficits in particular areas of the brain. Simple perceptual tasks, such as copying designs or telling the difference between words that sound alike, stymie the child who is "smart as a whip" in other respects, especially in finding ways to foil attempts to set limits.

The behavior itself, in many ways like that of the spoiled brat—annoying, intrusive, unpredictable, and persistent—shows an intensity suggestive of some serious impairment of the brain's inhibitory or braking ability. Activity level may appear to be excessively high, although measurement has tended to establish in most cases that the *choice* of activities is defective, not the level of activity. (Hence, the designation "hyperactive" is often a misnomer.) This restless, impulsive, inattentive pattern bears a close similarity to certain clinical conditions arising from known damage to the brain. For example, following epidemics of encephalitis in the 1920s, a syndrome known as *postencephalitic behavior disorders* emerged in the medical literature. Reasoning by analogy, some clinicians developed a hypothesis that many other forms of behavior and learning disorders might be due to impairment of brain function. Indeed, surveys of children with such problems show an excess of cases with problems in the pregnancy of the mother, in delivery, or in neonatal course—all capable of producing damage to the developing brain. The idea that there is a continuum of reproductive damage, ranging from minor at one end, to severe at the other, has received support from retrospective surveys (Pasamanick & Knobloch, 1960).

Another feature suggesting impairment of brain function is failure of the usual programs of treatment. Children who are anxious and maladjusted, whose families are overly protective or overly indulgent frequently respond well to brief psychotherapy or family counseling. Indeed, almost two-thirds of such children will improve no matter what form of intervention is applied (Eisenberg, Gilbert, Cytryn, & Molling, 1961; Levitt, 1957). However, less than one-third of hyperactive and behavior-problem children show improvement in response to a variety of treatments (Eisenberg & Conners, 1971). Although some improve spontaneously or respond to therapy, as a group they have a high risk for later psychopathology, social maladjustment, and vocational instability (Robins, 1979; Weiss, 1975; Weiss, Minde, Werry, Douglas, & Nemeth, 1971).

Earlier conceptions tended to rely on simple unitary notions of organic versus environmental causes for behavior problems. But there appear to be several distinctive etiologies which result in disturbances to the developing brain, and which interact with the child's environment and family milieu in complex chains. Recent work suggests that genetic background may be important in disposing the child to poor impulse control (Cantwell, 1972, 1975; Morrison & Stewart, 1971; Silver, 1971; Stewart & Morrison, 1973). Excess levels of lead in the blood (David, Clark, & Voeller, 1972; Silbergeld & Goldberg, 1973, 1974), high levels of smoking during pregnancy (Denson, Nanson, & McWatters, 1975), and perinatal abnormalities (Bernstein, Page, & Janicki, 1974) are among the causes of behavior and learning problems. Behavior problems in general cannot by any means be attributed solely to organic factors (Shaffer, 1973), but it is not at all unreasonable on the basis of present evidence to conclude that *one* of the results of brain dysfunction in children is a pattern of disinhibited behavior and specific learning disabilities.

Estimates have frequently placed the total number of children involved as high as 10% of the total child population, though more conservative estimates would suggest a figure of 1–3% of the population (Bosco & Robins, 1979). The sheer number of children places a heavy burden on the resources of clinics and schools, especially since the problems seem to be lifelong (unlike minor neurotic symptoms which appear to be quite transient for many children). But numbers alone do not give an adequate representation of the misery, despair, and disruption the children and their families experience. The child is frequently misunderstood as lazy or undisciplined and gradually finds himself friendless and isolated. He puzzles others and himself because of his obvious brightness and talents which nevertheless fail to protect him from academic failure. His social ineptitude and impulsiveness lead to scarred self-image and later maladjustment.

The families of such children are often severely strained by the frustrations imposed by a child who is constantly in need of control in order to prevent accidents and sometimes outright mayhem. Mothers are usually the unsung heroes who bear the brunt of the blame and responsibility, often coming to believe the common stereotypes which place all responsibility for deviant behavior in her lap. Despite an inner believe that *something* is different about this child, she will candidly admit that she has failed many times to be less than perfect in applying principles of reasonable child management. Mothers and fathers feel considerable guilt under the scrutiny of their all too human failings.

Parents are frequently ready to grasp at any straw. Some turn to theories based on vitamin deficiencies, resorting to treatments with megavitamins despite the lack of any scientific evidence for their efficacy. Others believe that some environmental hazard such as radiation or fluorescent lighting is to blame, again despite any hard evidence that such factors cause the problems (O'Leary, Rosenbaum, & Hughes, 1978). It is not surprising that a novel idea linking hyperactivity and food additives should receive immediate attention from parents of these children. The idea had a disarming simplicity and promises to reveal, at one fell swoop, both the causes of the problem and a simple method for its cure.

Dr. Feingold's Hypothesis

Dr. Ben F. Feingold was formerly a professor of Allergy at the Kaiser-Permanente Medical Center in San Francisco. He made his first presentation on food additives and hyperactive children in June 1973 in New York City at meetings of the American Medical Association. In September of that year he presented a similar report to a symposium on food health in London. By October, his reports had been picked up by Morton Mintz and publicized in the *Washington Post* (1973). Mintz's article and Feingold's presentation in London were reprinted in the Congressional Record of the United States Senate on October 30, 1973 at the request of Senator Glen Beall of Maryland. In June 1974, Feingold was interviewed on a New York radio program by Patricia McCann. These initial presentations aroused immediate interest among parents, professionals, and legislators.

Feingold's argument began with some troubling and indisputable facts. He pointed out first that according to government agencies (National Research Council and the Food Protection Committee of the United States National Science Foundation), there are over 2,700 "intentional" food additives. Preservatives, antioxidants, stabilizers, bleaches, sweeteners, food colors, flavors, and miscellaneous items in food are added to the natural ingredients for various purposes. He noted that of the 2,764 additives, flavors and colors constitute over 80% of the total. These items make no contribution to the nutritional value of the food; but, he added:

> [S]ince both flavor and color are very important determining factors in consumer acceptance, they are frequently interlinked. As a result, they play

identical and important roles in determining the economic success experienced in marketing food products, which in turn may subtly, yet at times very importantly influence the health and behavior of man. (Congressional Record, §19738)

Feingold went on to point out that the artificial colors have a long and murky record of being carcinogenic (cancer-causing), and that the list of acceptable colors is being revised constantly. According to him, these colors, largely derived from coal-tar derivatives, appear to have some "inherent potential for producing adverse effects" (Congressional Record, §1738).

The artificial flavors constitute by far the largest percentage of the total food additives (1,610 out of the 1,764). Feingold presented a table showing that seventeen different chemicals were present in artificial pineapple flavoring alone. He argued that modern food habits have changed in such a way that most people have come to rely on foods containing these artificial ingredients. He cited a report of the U.S. Department of Agriculture to the effect that only 3.3% of foods are prepared by the consumer from basic ingredients, and the rest come from convenience foods processed by the food industry. He suggested that the adverse reactions to these chemicals is actually much more widespread than is generally appreciated, largely because there is a lack of recognition that food additives are a relatively common cause of adverse clinical patterns.

Among the adverse reactions known to occur to these chemicals is a reaction to the yellow dye tartrazine (FD&C #5) in persons who are allergic to aspirin (acetylsalicylic acid). Despite the fact that tartrazine and aspirin are structurally unrelated, they seem to produce a common picture of adverse reaction. Feingold said that many patients with aspirin intolerance fail to improve until naturally occurring salicylates are removed from the diet. Foods thought to have high levels of natural salicylates include apricots, berries (blackberries, strawberries, raspberries, gooseberries), cherries, currants, grapes (including raisins, wines, and wine vinegar), nectarines, peaches, plums, prunes, almonds, apples, oranges, cucumbers, pickles, and tomatoes.

He then states his basic hypothesis:

Accordingly, on the basis of the clinical relationship between aspirin and tartrazine (FD&C Yellow No. 5), it was hypothesized that among the thousands of food colors and flavors incorporated into our food supply, there may be other additives, although unrelated chemically, which may induce adverse clinical responses. On the basis of this premise, the so-called salicylate-free diet was expanded to include not only all foods containing natural salicylates, but also all sources of artificial flavors and colors, with and without a salicylate radical. (Congressional Record, §19739)

Citing his own work (Feingold, 1968), he noted that practically all the bodily systems can be affected by flavors and colors: respiratory symptoms such as allergic rhinitis, nasal polyps, cough, laryngeal edema, and asthma; skin reactions such as pruritus, dermatographia, and angioedema; gastrointestinal symptoms such as constipation and buccal chancres; and neurologic symptoms such as headaches and "behavioral disturbances." Even though these symptoms usually occur singly, Feingold alleged that it is not uncommon for several reactions to occur at once.

But it was the behavioral disturbances that most caught Dr. Feingold's attention. He told a story about one of his adult female patients whom he was treating for aspirin intolerance with a salicylate-free and color-free diet. This 40-year-old woman was suffering with hives and swelling about the eyes. In studying her he learned that her hives and swelling cleared up when she was placed on the diet:

> But the remarkable thing was that about 10 days following that, I received a phone call from the chief of Psychiatry wanting to know what we had done because this woman had been in psychotherapy for two years without any response. . . . And as soon as we put her on a diet, it all cleared. So we started looking for more cases. We found a few adults, but we became more interested in the children who were having this. . . . (WOR Radio–TV reports, 1974)

Although these children came initially for treatment of hives or itching, the parents apparently would come back and say that not only the hives cleared up, but so had the children's behavior. Dr. Feingold found that, just as with his adult patients, he had fortuitously cleared up the behavior problems in children as well.

Dr. Feingold began to study the clinical picture of the hyperactive child and found, as we have noted above, that the problem is widespread, that the symptom pattern is quite variable and complex, and that treatments are often limited or ineffective. He then went on to point out that if one draws a graph showing the increase in dollar values for the production of artificial colors and soft drink beverages, it parallels exactly the increase in the incidence of hyperactivity and learning disorders among United States school children for the past ten years (Congressional Record, §19739).

From his clinical experience, Feingold felt he could extract several important features of the responses of children's behavior to the additive-free diet. First, there is a "rapid, dramatic change in behavior." This response is observed "within days" in children who may have had symptoms for many years. The sleep pattern improves, Second, drugs can be discontinued after two to three weeks of diet management. Third, improved scholastic achievement is "dramatic." It is a general principle of scientific study that if one can "turn the variable on and off at will," a causal relationship between the variable being manipulated and the response which occurs is demonstrated. Feingold argued that his ability to turn the symptoms of hyperkinesis and learning disorders off and on at will with the diet showed the causal relation of the substances in the diet and the behavior problems. He alleged that "in most cases a review of the diet diary reveals a larger than usual ingestion of artificial flavors and colors." He suggested that the pharmacologic properties of the colors and flavors could exert a kind of suppressive function on native abilities of the children, making them appear learning disabled or unmotivated in school (Congressional Record, §19739).

Another problem addressed by Feingold was why *all* children are not hyperkinetic, given the widespread dispersion of additives in the diet; and why only 50% of the children treated showed a favorable response to the diet. One obvious reason for the latter problem is noncompliance to the diet. Since so many items have to be excluded on the diet, occasional infractions are likely, and these presumably could set the child back in his progress.

Another reason why symptoms caused by food additives and improvement from an additive-free diet may not apply to all children is based on the principles of the science of pharmacogenetics. It is well known that the intermediary metabolism of many drug reactions is controlled by genetic factors, and that some individuals are highly susceptible to certain drugs because their metabolism has been altered by gene mutations or heredity. (Examples cited by Feingold are certain glucose deficiencies, the metabolism of the drug isoniazid, and an abnormal hemoglobin factor.) An important feature of such genetically controlled drug reactions is that they are often sex-linked. Feingold points to this as a possible explanation for the excess of males to females among hyperactive children. It is possible that the same genetic defect which leads to abnormal sensitivity to a class of drugs also produces some of the clinical features of behavior disturbance.

Finally, Feingold raises the possibility that some of the nonresponders to his diet are children who have already suffered irreversible damage, sustained *in utero* when the food additives crossed the placental barrier from the mother's blood supply and affected the growing nervous system of the child.

In summary, Feingold's hypothesis is a far-reaching speculation concerning a possibly genetically mediated sensitivity to artificial colors and/or flavors. Part of his reasoning is from analogy: the alarming increase in food additives parallels an apparent increase in learning and behavior disorders. The sheer magnitude of the numbers and amounts of chemicals consumed on the one hand, and the numbers of children who suffer from hyperkinesis and learning disabilities on the other seems to have caught his attention.

However, science and medicine usually insist that one go well beyond simple analogical inferences before serious attention can be paid to a hypothesis. It is true that there are large amounts of food additives in the diet, that they serve only cosmetic and nonnutritional purposes, that they are very widely distributed and consumed, and that there are also very large numbers of children who seem, inexplicably, neither to learn nor to develop socially in ways expected from their talents and backgrounds. But, one might argue, many fruits of technology, including pollution, industrial wastes in the water supply, low levels of ionizing radiation, crowding, and stress associated with urbanization have also spread geometrically, just as have food additives. It still needs to be demonstrated whether any of these phenomena are linked causally to behavior and learning disorders in children, no matter how plausible the connection might seem. Projections of the parallel rise of all these hazards is *consistent* with a causal relation, but certainly not *proof* of such a relationship. As Feingold says, the evidence for the causal relationship between food additives and hyperkinesis is more dependent on the success of his diet than on these plausibility arguments derived from simple analogy. It is, therefore, important to turn to the clinical evidence for the success of his diet.

Clinical Studies of the Feingold Diet

Feingold's original 25 cases were described in summary form at his 1974 presentation to the American Medical Association:

> The report indicated that a dramatic change in the behavioral pattern was observed within 1 to 4 weeks after instituting the diet. Children who were on medication with behavior modifying drugs, e.g., amphetamines, methylphenidate, tranquilizers, and antidepressants, etc. could discontinue the medication within a few days after initiating dietary management. In addition, in school age children a marked improvement in scholastic achievement was observed. The report indicated that any infraction of the diet, either deliberate or fortuitous induced a recurrence of the clinical pattern within 2 to 4 hours with persistence for 24 hours to 96 hours (4 days); in other words, we could turn the pattern on and off at will. (Feingold, 1974)

This same report goes on to describe five other clinical study programs which, at first glance, would seem to offer impressive confirmation of his ideas. In the first, 100 children were seen at the Kaiser-Permanente allergy department with a chief complaint of behavioral disturbance and learning disability. Of this group, 40 responded either dramatically or favorably. "Dramatic" was taken by Feingold to mean a "rapid and complete reversal of all signs and symptoms together with marked scholastic improvement" (Feingold, 1974). This report is somewhat puzzling since the figure of 40% dramatic and favorable response would seem to be considerably less than the previous report which found 68% improved. Forty percent is near the usual figures for placebo response in this group of children.

A second study involved 33 children at an elementary school. This study was organized and monitored by a graduate student in education. Only four of the children had a dramatic response and 15 showed a favorable response. No details about the study were reported by Dr. Feingold. The third study was a demonstration program funded by the California Department of Education. The population was a disadvantaged group of 25 children, all of whom were hyperkinetic. The state apparently provided all meals to these children because of their poor economic status. Four children responded dramatically and 12 had a favorable response. Seven of the nine nonresonders to the diet were said to follow the diet on an irregular basis. This study was described in a news release in July 1974, but apparently not reported elsewhere in any greater detail (Keithly, 1975).

A fourth study involved an unspecified number of children from the pediatrics department of the Santa Clara, California Kaiser-Permanente Medical Center. After a pediatric examination, nurse-practitioners managed the children and parents on the diet. Ten children had a "very favorable" response and four children were able, for the first time, to discontinue the use of drugs.

A fifth study was carried out at a residential care facility in Redwood Valley in northern California where the children were under 24-hour supervision. These children, ranging in age from 3–17 years, had diagnoses which included autism, emotional disturbance, and neurological handicap. Six of the eleven had a "very favorable" response, including improved sleeping habits, less frustration and unhappiness, improved attention span, and improved school work. The nonresponders included two children who were autistic.

The world-wide publicity generated by Dr. Feingold's book and early clinical reports led to the formation of the National Advisory Committee on Hyperkinesis and Food Additives under the sponsorship of the Nutrition Foundation. This foundation, though a food industry-supported educational and research organization, assembled a highly distinguished body of medical and behavioral scientists to review the Feingold hypothesis and data. Representatives were selected from the American Medical Association Council on Foods and Nutrition, the American Psychiatric Association, the American Association of Child Psychiatry, the Society of Toxicology, the Council for Exceptional Children, the American Alliance for Health and Physical Education, the Institute of Food Technologists, the American Society for Clinical Nutrition, the American Dietetic Association, the National Nutrition Consortium, the Life Sciences Research Office of the Federation of American Societies for Experimental Biology, and the Committee on Nutrition of the American Academy of Sciences. This impressive body was headed by a most distinguished scientist and scholar, Morris Lipton, M.D., Ph.D., the chairman of the department of psychiatry of the University of North Carolina.

This committee examined the clinical evidence and hypotheses offered by Dr. Feingold to that point. They concluded that "Dr. Feingold has not presented a specifically focused hypothesis." They made several trenchant criticisms. They pointed out that the degree of response approaches that reported from placebos. (A placebo is an inert substance used in drug studies, usually a sugar pill, to simulate all aspects of taking a drug except the actual chemical itself.) They noted that the total regimen of the diet program alters significantly the structure and dynamics of the families which could account for many of the improvements attributed to the diet. Dr. Feingold's charismatic personality and ability to generate positive expectations in patients might account for reductions of symptom severity. Parents and teachers were aware of the nature of the diet and thus "nonblind" to the treatment. Objective rating scales were not utilized in any of the studies. The ratings of improvement were global rather than specific.

The safety of the diet itself was called into question. It was argued that other nutrients or substances might be eliminated which could possibly account for the diet effects, and, in contrast to Feingold's claims that the diet could do no harm, the committee pointed out that restrictive dietary patterns might lead to deficiencies over a long period of time. It was recommended that studies include competent nutritional advice during use of any pediatric diet. Since there is no single class of substances which is eliminated, the obtained effects could come from many different sources, including needed vitamins.

A highly scathing critique was leveled by Dr. John Werry (1976), a distinguished researcher in pediatric psychopharmacology in an editorial discussing similar clinical studies:

> Dietary control . . . seems to have been at best rather cavalier, relying on unmonitored parent and child cooperation and self policing. How exact was the diet in meeting Feingold's actual hypotheses not his fantasies about what the foods contain? Where is the untreated matched control group or crossover design with half the subjects receiving dietary exclusions first and the other half the placebo diet first? What efforts were made to keep parents blind and thus avoid strong suggestive effects? We are all victims of "what we know, we see" effects and nowhere is this more true than in the management of children's behavior disorders. Further, the diet shifts the child from naughty to sick, gives his parents something other than punishment to administer and . . . causes the development of a type of camaraderie in the family which would be predicted to be highly therapeutic *independent of any specific dietary effect*. (p. 281)

Werry goes on to raise issues of accurate diagnosis, lack of appropriate rating scales, practice effects (the improvements from simply repeating a measure, independent of any treatment effect), and spontaneous remissions of symptoms which occur at a high rate in childhood disorders. Finally, he raises an issue of the ethics of introducing treatments into medicine before they have been shown to be efficacious and safe. He raises the possibility of lack of safety from potential vitamin reduction during the diet. Regarding efficacy of the diet, he says:

> I personally feel there is no greater breach of medical ethics than that of foisting a potentially worthless or dangerous treatment on to a credulous public. Theirs may be the right to believe in magic and panaceas but ours as a profes-

sion is to act responsibly, cautiously and scientifically, though not prejudi-
cially. Why should all my U. S. colleagues in pediatric psychopharmacology
research, no more than a handful, have to drop their work to show to a
clamouring public that Feingold's hypothesis is or is not correct and is or is
not safe? Surely the obligation is his before he announces it to the public? (p.
282)

One obvious answer in this case is that Dr. Feingold at 75 is a man in
a hurry. He once told me while we were on a radio program together, "I
don't have time for sacred cows of science, the double-blind placebo-
controlled trials." I raised many of the scientific issues his other critics
had raised, but he said these were irrelevant because he had already
proven the hypothesis by his ability to turn the symptoms off and on
with the diet. Rather than support these assertions with laborious and
time-consuming studies, he preferred to take his message directly to the
consumers.

One such occasion was the national meeting of the Association for
Children with Learning Disabilities, held in New York City in 1975. I was
amazed at the size of the audience which crammed every corner of a
large ballroom; they had come to hear Dr. Feingold and his new theory.
This association is composed of parents and professionals with a common interest in learning disabilities, and news of Dr. Feingold's theory
had aroused very high interest. Two things stood out as impressive to
me.

First, there was the tremendous receptivity of the audience to this
new point of view. Here was an idea that promised new understanding,
both of cause and cure. For most of the audience—whether the teachers
who know how slow and painfully progress comes with such children,
or parents who must endure daily the frustrations of the child who
"marches to a different drummer"—there was eagerness and hope. On
the one hand, it appealed to what everyone knew or suspected, that the
food supply, like modern life in general, is fraught with hazards
wrought by new technology. On the other hand, these dangers are unnecessary, indulged in for cosmetic reasons only, and designed around
marketing rather than nutritional considerations. The idea that "natural
is best" is something most parents accept readily. Everyday the news-
papers provide fresh awareness of the growing hazards of pesticides,
heavy metals in food and water, radiation, and other offshoots of industrial and scientific technology. Feingold's message needed no support in
that respect.

Moreover, Dr. Feingold's presentation left the audience with a general sense of scientific rigor and medical plausibility. He first spoke with undoubted authority of processes he knew well [having himself made important contributions to allergy research and having written a highly regarded textbook on allergy (Feingold, 1973)]. He started with facts and figures taken from reliable sources regarding the chemical constituents in foods. He pointed to well-known and firmly established associations between nutrition and disease and between nutrition and brain function. He merely repeated indictments made by many that the food industry and the FDA were lax in screening food additives. He pointed out the known carcinogenic and teratogenic hazards of the coal-tar derivatives in food colors. As a pediatrician, he spoke articulately as an advocate for the newborn and young children who could find themselves poisoned before having any choice over what they eat. The audience responded with thunderous endorsement to a message that seemed to be based on fact, that offered hope, and that gave a concrete solution to some very difficult problems.

Dr. Feingold smiled cheerfully as he sat down, turned to me and said, "They're all yours, Keith." I felt like the messenger bringing bad news to the king, for my message was not one of hope, but one of caution. I presented some preliminary data which showed that although some children *seemed* to improve on Feingold's diet, others did not. I presented the standard arguments for the necessity of control groups, blind ratings, and placebos. The audience was attentive but hostile; they were not interested in "scientific paraphernalia." Any failure of the treatment could be explained readily as the result of noncompliance to the diet; any successes could be attributed to the removal of food additives. Parents were impressed at the *dramatic* nature of the alterations in their children and refused to believe that such changes could be due to "mere" placebo effects. I pointed out that the megavitamin approach, discredited by professional opinion, used similar arguments about nutrition and brain function and, like Feingold, went beyond the data to speculations about treatment. I made the point that placebo effects may be dramatic and powerful, and that there is a typical natural history of new drugs: at first, euphorically accepted as promising, uncontrolled studies show a high rate of success. As experience deepens, skeptics begin to doubt the initial reports, and finally, when controlled studies are done, the effects disappear entirely. The tongue-in-cheek maxim is, "Use the new drug quickly while it still works."

Some treatments (like the "creepy-crawly" treatment of brain damage proposed by the Doman–Delacato school) involve such rigorous demands that failures can be attributed to lack of compliance with the regimen. Usually these treatments are extremely difficult to verify in other people's hands, leaving its authors a clear field for their claims. Often, these new treatments are denied appropriate testing by their authors who prefer instead to set up private clinical practice for those who have lost hope from "standard" medical practice. A conspiracy theory between big business and the medical profession is alleged to account for the fact that the treatment is not generally accepted. The medical and scientific establishments are painted as conservative self-interested bodies who practice trade union and guildlike policies to keep down the competition. Because of the chronic, unremitting nature, the complexity of symptomatology, and the high prevalence, learning and behavior disorders are particularly subject to the prey of therapeutic entrepreneurs.

Verifying new therapeutic claims is difficult because scientific inquiry is a slow and painstaking process. Within the fields of psychiatry and behavioral science, even basic methodologies are not agreed upon and clinicians are often suspicious that scientific studies may not only delay, but obscure the true processes.

Sometimes the attempt to impose formal and rigorous standards succeeds in hiding the very phenomena one wants to study. I had myself argued previously:

> Careful selection of homogeneous samples, randomization of assignments to drug and placebo, counterbalancing of treatment or test conditions, objective, valid and reliable measurements, and a number of other common scientific considerations may be desirable but impractical within the treatment context. Moreover, it is likely that some degree of control is actually sacrificed in drug trials when these considerations are met.
>
> For example, the effort to use groups large enough for statistical analyses, with randomized assignments to treatment conditions, may well lead to a trial conducted on a heterogeneous sample, with some real drug effects being canceled by negative effects, or washed out by a number of nonreactors included in the sample. In this instance, a careful "clinical" trial that is "uncontrolled" may, in fact, be superior to an ostensibly tighter design. Clinicians may be able to group patients in small but homogeneous groups, and they may then detect improvement or change in areas that would be insensitive to "objective" measurement. (Conners & Werry, 1979, p. 337)

These are legitimate difficulties in the way of testing new clinical ideas rigorously, but they do not justify abandoning efforts to carry out controlled studies. Good clinical ideas need to be tested with as many controls as possible, because experience has shown how frequently plausible treatments turn out to be ineffective or due entirely to nonspecific effects. These nonspecific effects include, most importantly, what is known as the *placebo phenomenon*. In actuality, we know little of how placebos work. But that they do indeed work in reducing symptoms in many diseases, almost magically in some instances, attests to the necessity of evaluating their role with any new treatment. Although Dr. Feingold has said on many occasions that he doesn't care how his diet achieves its effects, even if it should turn out to be through the placebo process, there are solid reasons why it does make a difference.

First, as we have seen, there are questions about the safety of any new diet with rapidly growing children. Second, placebo effects are usually transient (as are most treatments which merely suppress symptoms rather than eliminate the underlying cause of the disease). Third, erroneous inferences about causal relationships leads to unnecessary and expensive explorations in areas that waste valuable scientific resources. And finally, the disappointment over time wasted in treating the real cause of the problem leaves parents and professionals frustrated and angry, and the children further behind in their struggle to survive.

Clinical Case Trials

Dr. Feingold based his argument partially on his own clinical experience with hyperactive children. Typically, a mother would be given instructions for following the diet and would return a few weeks later and make an informal report on the child's progress. Feingold describes the nature of these changes as follows:

> The experience to date indicates that approximately 50 percent of children with H–LD respond to strict elimination diets. Loss of hyperactivity, aggression, and impulsiveness are the initial changes observed. This is soon followed by improvement in muscular coordination as indicated by improved writing and drawing abilities, greater facility with speech, and loss of clumsiness . . . scholastic achievement improves rapidly. (Feingold, 1975a)

These clinical observations are typical of the initial stages of scientific observation. A clinical problem is identified, an intervention is made, and the degree of progress following the treatment is determined. The observations are informal, and documentation at this stage is minimal. The study is logically a simple A–B treatment design: Treatment A is given, and effect B is measured. Impressive effects at this stage are sufficient to encourage a more carefully controlled trial. Failure at this level has many possible reasons but usually does not warrant a careful and expensive controlled trial.

We were first interested in seeing the bare phenomenon for ourselves with very few controls. If the simple A–B design is itself not replicable, then there is no real phenomenon to study. We conducted several such trials on hyperactive children who had been diagnosed and treated previously in our clinic. These individual trials consisted simply of observing the child for several weeks, placing him on the Feingold diet for several weeks, removing the diet, observing for several more weeks, and repeating the process. This A–B–A–B "off–on–off–on" trial is essentially like the original case studies in that it is uncontrolled. On the other hand, it does use patients who are formally diagnosed as hyperactive (thus ruling out anxious or mild behavior problems who usually get better spontaneously at a high rate), and careful instructions and records regarding compliance to the diet are maintained. None of these niceties were included in Feingold's original reports.

In the first case, a child was observed with ratings made by the mother using a simple checklist of behavior. This ten-item scale has been used in many carefully controlled drug trials and has proven to be reasonably reliable, sensitive to treatment effects, and capable of significantly distinguishing between hyperactive and normal children (Conners, 1973a; Kupietz, Bialer, & Winsberg, 1972; Werry, Sprague, & Cohen, 1975). The usual entry score for children in our studies is a score of 15. This score is exceeded by only 2.5% of the normal population. The scale is shown in Figure 1.

After being observed for three days of the first week, the child was placed on the diet and observed for four weeks, taken off the diet for one week, and placed on the diet again. The results are shown in Figure 2. Note that during the baseline, the child showed a mild improvement but by the second week of the diet showed a dramatic lessening of symptoms reported by the mother. When a normal diet was reinstituted, the report of symptoms promptly increased. The symptoms diminished somewhat during the second baseline, but continued to decline even further during readministration of the Feingold diet. This pattern of response certainly appears to be in accord with the clinical report of Dr. Feingold.

A second case illustrates an opportunity for some additional observations. This child (Figure 3) has a high level of symptomatology which remains unchanged during the baseline week.

OFFICE USE
Patient No.
Study No.

ABBREVIATED PARENT QUESTIONNAIRE

PATIENT NAME _____

PARENT'S OBSERVATIONS

Information obtained _____ by _____
Month   Day   Year

Observation	Degree of Activity			
	Not at all	Just a little	Pretty much	Very much
1. Restless or overactive				
2. Excitable, impulsive				
3. Disturbs other children				
4. Fails to finish things he starts - short attention span				
5. Constantly fidgeting				
6. Inattentive, easily distracted				
7. Demands must be met immediately - easily frustrated				
8. Cries often and easily				
9. Mood changes quickly and drastically				
10. Temper outbursts, explosive and unpredictable behavior				

Comments:

_____

_____

_____

_____

_____

Figure 1. Ten-item abbreviated rating scale. Each item is scored 0, 1, 2, or 3. Referred to in text as Abbreviated Parent Questionnaire (APQ). The same items are used by teachers.

Again, there is a dramatic lessening of symptoms during the diet phases, but contrary to expectation, there is no worsening during the repeat of the baseline phase. The diet diary revealed that at week six there were several violations of the diet, and accordingly, the mother reported a worsening of behavior. When we questioned the mother why the child did not get worse during the second baseline phase, she admitted somewhat sheepishly that the improvement had been so marvelous that she couldn't really take the boy off the diet as she had agreed to do.

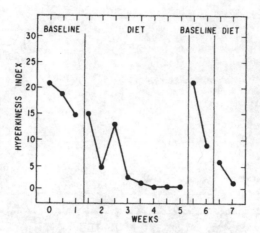

Figure 2. Open diet trial showing reversal of symptoms while on baseline (normal diet) and an additive-free diet. Lower score indicates fewer symptoms.

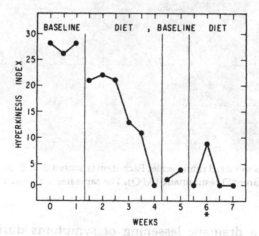

Figure 3. Open diet trial showing failure of symptoms to return during second baseline (when mother decided to keep child on additive-free diet). Note worsening of symptoms during period (week 6) when child violated diet with many infractions.

Thus, the failure to worsen on the presumed baseline would be understandable.

A third case presents still another compelling vignette. In this case (Figure 4) the child showed a worsening of behavior during the latter part of both diet phases. The mother attributed this worsening to the fact that the child was receiving an oral antibiotic for a sore throat. Dr. Feingold has pointed out frequently (Feingold, 1975a,b) that the artificial coloring in vitamin supplements and drug tablets might itself be the cause of an adverse reaction. (Of course, one might equally surmise that the increase in behavior problems noted by the mother during this period could be secondary to the effects of the sore throat on the child's behavior.)

In any case, illustrations of clinical effects such as these are highly persuasive to both parents and clinician alike. The mothers typically report the improvement in incredulous tones, and the clinician basks in the warmth of a grateful parent, modestly taking the credit for the delightful outcome. From a scientific point of view, however, it is obvious that the mother knows when the child is being treated and when he is not, and that she is carrying out a rather subjective form of observation. Even if the parent is scrupulously honest and objective, merely knowing that the child is receiving a new and hopeful treatment in the hands of a respectable and authoritative doctor could bias the direction of rating. In

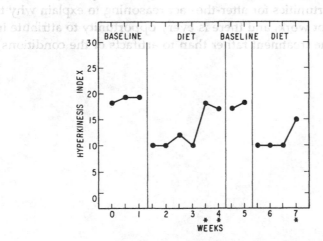

Figure 4. Open diet trial. Asterisks indicate increase of symptoms when oral antibiotics were taken.

addition, the diet compels the parent and child to work more closely together, excitement is generated about a new and positive possibility for better relations in the family, and the child may be eager to confirm all the nice things that the doctor and parent say can happen.

There is also no doubt from these clinical trials that the child is a very active participant. Almost all the children we have studied have been eager to cooperate. The child will frequently and proudly report how he resisted the temptation of treats not allowed on the diet. A child who is eager to maintain the good feelings of all the interested parties, and who actively tries to do what is expected might change enough to account for some of his mother's perception of improvement during the diet phase, particularly since the child is well aware, under these conditions, of what the diet is supposed to do.

Although these open studies are very persuasive psychologically to the clinician and parent, their limitations are obvious from the point of view of proving a hypothesis. The dramatic way in which the diet affects perceived behavior is certainly apparent from these and many similar case studies. It is not surprising that Dr. Feingold would have become very impressed with the potential efficacy of this treatment given a repetition of this experience over many cases. Without the proper controls, however, such information is merely suggestive. To an unsuspecting clinician, such case studies can seem quite compelling. There are numerous opportunities for after-the-fact reasoning to explain why the treatment did not work, and there is every opportunity to attribute improvement to the treatment rather than to artifacts of the conditions of measurement.

CONTROLLED DIET TRIALS

Design of the First Controlled Trial

The simplest form of controlled clinical trial is one which duplicates exactly the conditions of the original observations, and which ensures that only the variable of interest is manipulated. Typically, one group receives the A–B form of treatment and follow-up, and another group receives a treatment that is similar in every respect except that the active ingredient is missing. In human clinical trials, this control group receives all the nonspecific ingredients that might account for the apparent improvement, such as expectancies of being helped, spontaneous improvement, temporary improvement due to alterations in routine, and extra attention. Another important factor controlled in such designs is the bias of the observer who typically is not blind to the treatment, who may be biased in its favor, and who honestly sees improvement.

In the most common study, the patient is assigned randomly to one of two treatments, such as a drug or a placebo. The doctor is then blind to which treatment the patient receives. If the patient is also blind, then the trial is said to be double-blind. Such a study usually requires large numbers of subjects if there is much variation with respect to the effects being measured. The measurement of behavior patterns in children and adults is notoriously variable and affected easily by biases of an observer. To help eliminate this variability between subjects which might obscure treatment effects, it is common to have each subject serve as his own control; that is, receive both the active and the placebo treatments.

This design, however, creates additional complications. If the subject receives two treatments, they might influence one another; observers might change their standards across the two conditions; or the passage of time between treatments could allow changes to occur which might be attributed to the active ingredient. It is usual, therefore, to counterbalance the treatments, with one group receiving the active treatment first, and one receiving the placebo treatment first. This counterbalanced crossover design allows one to compare the patient under both treatments, and to evaluate any sequence or order effects in the observations. This is the design we elected to use in our first study of the Feingold diet.

Selection of Patients

One of the important limitations of Feingold's original clinical observations was that he used no standard method of selecting and describing his patients. It is now a well-accepted fact that behavior problems in children stem from many different causes. Some children are fearful and anxious and therefore concentrate poorly, giving them the appearance of being hyperactive. Such children often have problems of adjustment in school or at home which disappear rapidly once conditions change for the better. A group of such children will, therefore, usually have a high rate of spontaneous improvement. If the parent or teacher seeks help in changing problems in these environments, improvement can occur in a very short time and the therapist gets credit for a change that would occur by itself. Typically, two-thirds of anxious problem children improve with brief counseling or the passage of time (Eisenberg, Gilbert, Cytryn, & Molling, 1961).

Other children are hyperactive because of early birth trauma, head injuries, delayed maturation, or genetic factors. There are many different patterns of neuropsychological deficit in children whose main complaints are that they are hyperactive (Conners, 1973b). Only some of these may be sensitive to food additives. Food additives might be like low levels of lead in the blood which appear to cause hyperactivity mainly in those "pure cases" of children free of other causes for their problem (David, Hoffman, Sverd, & Clark, 1977). In any case, it is important to document the *type* of child whom one chooses to call hyperactive.

Unfortunately, there are no universally agreed upon rules for selecting truly hyperactive children. The best available method seems to be a careful review of the child's birth, developmental history, and family background, followed by an examination of his intellectual skills, social behavior, and school behavior. This selection process usually starts with an interview with the parent and takes about one day. During this interview, one assesses the quality of the parents' relationship with each other and their attitude toward the child. The parents are questioned carefully regarding the attitudes and circumstances surrounding the birth of the child. The child's early sleeping, feeding, and motor patterns are queried. His early school years and present behavior are discussed in depth.

After the interview, the clinician usually spends an hour with the child, informally putting him at ease so as to draw him out about his feelings about home, school, and himself as well as to observe his behavior and reactions during the interview itself. A series of brief tests of coordination, balance, rhythm, and other neurodevelopmental skills are performed, and the child is given a physical examination. A psychologist usually administers achievement and intelligence tests and observes the child while he attempts to solve problems or perform reading, spelling, and arithmetic tasks. Parents and teachers fill out checklists of the child's typical behavior.

On the basis of all this information, a judgment is made as to whether the child fits into the category of the hyperactive child, or more appropriately falls into some other category. Other categories might include mental retardation if the child's intelligence is low, anxiety or withdrawing reaction, learning disorder, or severe reactions such as childhood schizophrenia or psychosis. Although this is a subjective process, clinicians have come to recognize the most prominent features of each category, and it is only by a careful review of all the data that this determination can be made. It is helpful to have formal criteria such as a cutoff score on our rating scale, but these can be deceptive if used by themselves.

In the present study, children and their families were examined by a clinical psychologist and a child psychiatrist. In order to be accepted for the study, the child had to meet the following criteria: be between 6 years and 12 years, eleven months of age; have an IQ of 85 or above (i.e., low average or better) and meet criteria for the syndrome as outlined by a National Institute of Mental Health (NIMH; 1973) working group:

NECESSARY AND SUFFICIENT SYMPTOMS

Hyperactivity—with a high and conspicuous level of gross motor activity (locomotion; or "rump" hyperactivity when seated, i.e., squirming, changing position and getting up and down frequently; but not finger–hand-twisting, picking or other small muscle activity) occurring across environments in situations in which sedentary or quiet behavior is appropriate for age.

and

Disorder of attention—with higher distractability and shorter attention span than appropriate for chronological age (*not* mental age), especially in school or group situations.

SYMPTOMS COMMONLY ASSOCIATED BUT NOT SUFFICIENT FOR DIAGNOSIS

Poorly integrated and labile behavior, which gives the impression of immaturity and of uneven but generally inadequate abilities.

Extremely variable relation to adults (including examiner), with rapid fluctuation from attempts at compliance to silly clowning, boisterous, mischievous or impertinent behavior, clinging and demanding behavior, and/or angry or sullen negativism.

Labile affect. React with excessive irritability to any situation interpreted as rejecting, demanding or restricting, with angry, suspicious, anxious, unhappy, and silly clowning responses, often associated with gross motor discharge, tantrums, destructive, or aggressive behavior.

Speech is often sparse and unelaborated, with a tendency to evade emotionally charged material.

Fantasy is usually expressed more clearly in play; concerned with movement and aggression, diffuse fears of retaliation and loss of love.

Motility usually variable, impulsive and poorly coordinated. Movements are relatively undifferentiated for age; have difficulty suppressing gross body movement when attempting isolated, finely coordinated finger, hand, or arm movements. Body manipulation relatively uninhibited for age; chewing, sucking, nose picking, masturbation.

Unable to conform to demands of a group situation with peers; often become scapegoats and/or participate peripherally by provocative, silly, teasing, aggressive, quarrelsome behavior; usually considered "babies" and "pests" by peers.

Adults usually consider them immature, demanding, difficult to manage. Have chronic and recurring difficulties in adapting to age-appropriate social and educational demands.

DISQUALIFIERS

Psychosis—If so permeated by autistic preoccupations or thought disorder, as defined under schizophrenia, as to necessitate a diagnosis of psychosis, then classify as *Childhood Schizophrenia. Expressed preoccupation with anxiety and sadness* which is pervasive, *NOT* transient.

Unsocialized Aggressive Reaction with organized behavior pattern.

In order to arrive at this diagnosis, the psychiatrist first conducts a standardized interview with the child as described by Rutter and Graham (1968). He then rates the child on 63 items or scales known as the Children's Psychiatric Rating Scale (CPRS), which is presented in Appendix 1 (this vol.). He also performs a brief physical examination and neurological examination. All this information is then integrated with the results of the parent interview to form an overall judgment of type and severity of the condition. In addition to these criteria, our patient had to be judged to be "moderately ill" on a seven-point scale going from normal to extremely severe, and have a score of 15 or greater on the ten-item or Abbreviated Parent Scale (APQ).

Procedure

Once the assessment was completed and the child was judged to fit the selection criteria, parents and teachers independently completed biweekly questionnaires regarding the child's current behavior, utilizing the ten-item symptom scale over a two-week period (pretreatment period). If the child was receiving stimulant medication, it was discontinued at the end of the pretreatment period. All children were then observed for another two weeks (baseline period).

At this point, the children were designated randomly to receive either the experimental Feingold diet (referred to as the KP diet, or Kaiser-Permanente Diet, because of Dr. Feingold's association with that foundation); or they were assigned to the control diet. This was a very important phase of the study because we wished the parents to believe that either diet might be effective so as to control for differences which might reflect their expectations or beliefs about the relative efficacy of the two diets.

The parents met with the staff nutritionist, who explained that we wished to see which diet might be best for their child, and that their child would have the opportunity of being treated with both. We stressed that they should observe the child as objectively as possible and follow the diet procedures carefully.

The control diet was devised with the following goals in mind: (1) the diet should involve the same degree of time in preparation, shopping, and monitoring as the KP diet; (2) the items in the control diet should be drawn from the same food groupings and categories, where possible, as the KP diet; (3) the two diets should be nonoverlapping; i.e., items on the control diet should allow for eating items excluded on the KP diet, and vice versa; (4) the control diet should be as palatable and as easy to follow as the KP diet; and (5) the control diet should appear plausible and reasonable as a possibly effective treatment. The words "KP diet," "Feingold diet," or "control diet" were never used with the parents. Instead, parents were told that their child would try two experimental diets, that either might produce improvement, and that it was necessary to have both diets to compare with each other. They were told we were studying dietary factors in behavior problems, and that there might be a number of separate food items that could cause behavioral difficulties, and that only by comparing different approaches systematically could we be sure which diet might be effective for their child.

Assessment of Prior Dietary Status

The nutritionist interviewed the parents and collected information on the child's food habits, likes and dislikes, allergies, and any restrictions or medical problems with foods. They were asked to recall everything the child had eaten during the previous day, including the time, place, methods of preparation, and amounts. Food models were used to assist in assessing the amount. The parents were then given a list of foods and asked to record the number of times or frequency each food was usually consumed by their child during a typical week (see Appendix 2, p. 121).

Monitoring Compliance and Nutrient Intake

During the 12 weeks of the study, parents kept diet records of everything their child consumed during six days of each month. The time, place, type of food, and a description of the method of preparation and amount consumed were recorded on a form. These methods, in which records of 3–7 days of diet behavior are recorded have been shown to give a reliable enough estimate of food intake for calculating the nutrient intake (Beal, 1967; Chalmers, Clayton, Gates, Tucker, Wertz, Young, & Foster, 1952). These records were later coded and analyzed for calories, protein, carbohydrates, fat, calcium, iron, vitamin A, thiamin, riboflavin, niacin, and vitamin C. The averages for each of the three months could then be calculated to determine whether the three phases of the program (baseline, first diet, second diet) differed in nutritional characteristics.

The Experimental and Control Diets

The nutritionist spent as much time with the parent(s) as necessary to clarify the study procedures. At first, the number of items to be attended appear to be overwhelming to the parent and child. There are numerous details and records to keep. Among these is a checklist of infractions of the diet. Both parents and child are encouraged to maintain a strict compliance with the diet, but we acknowledge that perfection will be impossible and that at times there might be inadvertent or even willful lapses. Nevertheless, they are encouraged to report these without fear of censure.

The composition of the diets and the procedures to be followed were given to the parents in a small booklet. At this time the parents also signed an "informed consent." This is a procedure required by special committees which review human experimentation in order to protect the interests and privacy of the participants. This statement is actually an important part of the conditions of the experiment since it sets the expectations, defines the contract between the experimenter and the family participants, and conveys other information about the study. The dietary procedures, the study procedures, and the informed consent statement are reproduced in Appendix 2.

Monitoring the Effects of the Diets

When the family was ready to begin the study, a random number table was used to assign the child to one or the other of the two diets. The parents were given the appropriate diet information and a supply of rating forms. They were instructed to rate the child's behavior on Saturday of each week for the preceding week. Teachers were sent similar forms and instructed to follow the same procedure, except they rated only the five-day period ending on Friday. The parents were instructed to contact us by telephone to discuss any problems with the diets or procedures. A record of these calls was kept in order to determine whether there was any difference in the perceived difficulty of the two diets.

At the end of the four-week period, the parents and child returned for another office visit. The project coordinator met with the parents and instructed them that an interviewer would ask about the child's progress during the past month. They were instructed not to reveal to which diet the child had been assigned, so that the interviewer could judge progress solely on the behavior changes reported. Similarly, the interviewer greeted the parents and then instructed them not to reveal any details about the particular foods.

In the assessment interview, the parents' view of the overall changes, changes in physical function, relations of the child with peers and family, and the child's reactions to the diet were discussed. At the end of this interview, the interviewer recorded a judgment of improvement on a four point scale: no change, mild improvement, moderate improvement, or excellent improvement. This Clinical Global Impression Scale (CGI) seems crude and unreliable, but such overall judgments are often among the most sensitive indicators of change in double-blind studies with drugs and can be surprisingly reliable. At the end of each phase of the study, the parents and teachers filled out longer checklists of behavior (39 items for the teacher and 93 items for the parent). Embedded in these longer ratings are the ten items from the abbreviated scale which they had filled out each week during the study.

After the first diet was completed, the parents were given the instructions for the second diet, and the same procedures as before were followed for another month. In the final interview, the parental perception of the two diets was reviewed and they were discharged. The parents were told that we would contact them later to determine whether the child's behavior had changed, and they were encouraged to follow whichever diet they felt was most beneficial.

Comparison of Results from the Two Diets

Clinical Global Impressions

Table I shows the improvement ratings for the 15 children so that the body of the table indicates how many children received a given rating under each of the diet conditions. For example, the table shows

Table I. Clinical Global Impressions Scale Results

		FEINGOLD DIET				
		0	+1	+2	+3	
C	0	4	4	3	0	11
O D						
N I +1		1	1	0	1	3
T E +2		0	0	1	0	1
R T +3		0	0	0	0	0
O						
L		5	5	4	1	15

Note: 0 = Unchanged or worse; + 1 = minimal improvement; +2 = moderate improvement; +3 = marked improvement.
Wilcoxon Matched-Pairs Signed-Ranks Test: $p = 0.01$ (one-tailed).

that three children received + 2 ratings while on the Feingold diet and a zero rating while on the control diet. Only one child was rated as showing the maximum improvement on the Feingold diet, and he showed a + 1 (slight) improvement while on the control diet. One child showed a moderate + 2 rating of improvement on both diets. Important, perhaps, is the fact that no children showed a + 3 improvement on the control diet. The probability that these paired ratings could have arisen by chance is less than one in a hundred. If we take a conservative point of view and count as true responders only those children who improved slightly on the control diet and who also improved at least moderately on the Feingold diet, we see that only four of the 15 children can be considered responders.

Parent and Teacher Reports

The average scores for the children during the pretreatment, baseline, and diet phases are shown in Figures 5 and 6. These scores combine the data from the patients who got the diets in different sequences. (By chance, nine of the subjects received the KP diet first, and six received the control diet first.) To examine the significance of these findings, each child's ratings under the two diets were corrected for the starting point by subtracting out their baseline scores. These baseline-corrected scores were then tested for significance by a procedure known as *analysis of variance*. The results indicated that the teachers rated behavior as significantly improved when the child shifted from the baseline period to the KP diet, and that behavior was rated significantly worse when the child was on the control diet compared to the KP diet (probabilities are 5 in 100 and 5 in 1,000, respectively).

The parents, on the other hand, rated the children as significantly improved when they shifted from baseline to the KP diet, but they did not consistently detect a difference in behavior between the two diets. Sometimes, however, the parent and teacher were in striking agreement as may be seen in Figures 7 and 8. There appears to be a considerable concordance between the teacher and parent ratings for these two children, as though both sets of observers were noticing the same fluctuations in behavior over time.

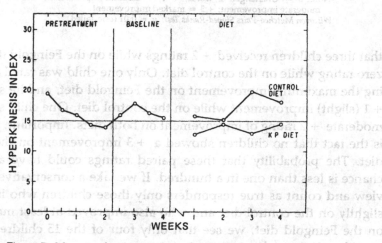

Figure 5. Mean teacher rating scores for 15 children averaged across treatment orders.

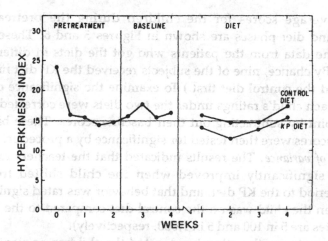

Figure 6. Mean parent rating scores for 15 children averaged across treatment orders.

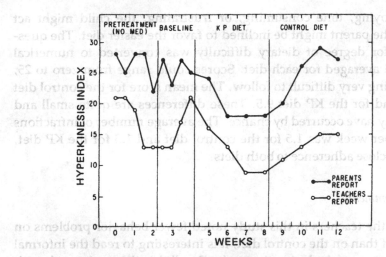

Figure 7. Data for an individual child showing concordance of parent and teacher ratings. Child received Feingold diet (KP diet) first, control diet second.

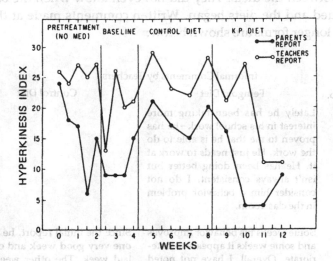

Figure 8. Data for an individual child who received control diet followed by KP diet.

Ease of Following the Diets

It is important to know whether the two diets were comparable with respect to problems encountered in following them. If one regimen was more annoying, time consuming, or frustrating, the child might act worse or the parent might be inclined to favor the easier diet. The questionnaire for degree of dietary difficulty was converted to numerical scores and averaged for each diet. Scores could range from zero to 25, with 25 being very difficult to follow. The mean score for the control diet was 8.3 and for the KP diet 9.5. These differences are quite small and could easily have occurred by chance. The average number of infractions reported per week was 1.5 for the control diet and 1.3 for the KP diet, indicating close adherence to both diets.

Informal Comments

Since the teachers in this study noted fewer behavior problems on the KP diet than on the control diet, it is interesting to read the informal comments they made during the study. Recall that the teachers, though aware that the children were in a diet study, presumably did not know there were two diets involved and did not know when the crossover occurred between the diets. They did not even know when the baseline period ended and the diets began. Written comments made at the bottom of the longer forms are shown below:

Informal Comments by Teachers

Patient No.	Feingold Diet	Control Diet
004	Lately he has been taking more interest in his school work—he has proven to me that he is able to do the work. He just needs to work at it. He has been doing better but isn't always consistent. I do not consider him a behavior problem in the classroom.	
008	Some weeks his behavior improves and some weeks it appears to deteriorate. Overall, I have not noted any significant improvement in his behavior. At times I feel that his overall behavior may be a little worse.	Since the initial report, he has had one very good week and one very bad week. The other weeks have been the same or generally the same.
010	The biggest change has been his almost happy acceptance of warm gestures from me, such as an arm	

Patient No.	Feingold Diet	Control Diet
	around his shoulder. Previously, he would recoil from a mere pat. While he still tends to deny responsibility for misbehavior, he is less shrill or excitable about it. All in all he seems to be a little more "put together."	
014	He has acted pretty much the same since he has been taken off the medication.	None
021	While he seems less to be groggy—or medicated—he is also less able to concentrate or use any controls to facilitate learning. He absolutely requires a one-on-one to get any sensible responses from him, though more often than not you get an irrelevant comment. Frequently he indicated he was really trying but the results were minimal.	There are no positive changes. If anything he is demonstrating less rational behavior than before.
022	He seems to be an overly active child; he is doing more work in school than he was doing previously.	At times still fidgets, is overactive and disturbs the other children. However, he has improved in the above areas. I don't hear about him from the other children. I only hear a few.
023	He has been somewhat more settled and has not been popping in and out of his seat as often as before. However, he has not been any quieter or more attentive.	He has been less fidgety and better able to sit in a chair than earlier in the year. He does not turn around as often and is somewhat more calm.
027	Less radical mood swings, less or shorter periods of depression, more consistent at "following directions" or working on a given task with less supervision.	Attention span much shorter; more frequent crying without apparent reason; aimless running.
028	Actually during this period his behavior became more erratic and at times worse than had previously been observed except during the aforementioned period of time. Academically, he continued to make progress as he has all year.	His behavior has not improved since my first report was filled out. He has his good and bad days. If anything, he has been having more consecutive bad days.

Patient No.	Feingold Diet	Control Diet
029	The pattern is much the same, except that behavior is worse. There is much more activity, much more evidence of frustration.	He still distracts children from work, even when he's involved in an art project. However, there is not so much interference with others at playtime. There is less withdrawing, more talking back. For a while he was fascinated with the words "penis" and "bum" and often drew pictures of these.
030	Physical reactions against other children have dropped almost completely. He is able to take correction without gross reactions. Picking up loose objects when requested. Calmer in class—more manageable in group—responds better to praise—seems interested in making friends.	His physical aggression has dropped. He rarely hits other children. Last 10 days prior to March 5/8 seemed almost normal.
033	When given a punishment assignment, she appears very quiet, but will cooperate.	Seems more excitable.
036	He had one very good week the last of April, but with that exception he has been much more uncontrollable than when on medication. He has begun to steal from other children and hide what he takes.	At the previous report he would respond to behavior modification techniques being used, but he has lost touch with reality and only comes from his fantasy world on occasion, to strike out violently. He rolls about on the floor and when an attempt is made to him he bites, hits, or punches.

Not all teachers made comments and when they did so, it was spontaneously.

Several of the comments appear to show improved behavior or school work while on the Feingold diet, but note also that in many cases similar improvements are noted on the control diet when the child is presumably being exposed to food additives which ought to make him worse. In some cases the behavior is worse on the KP diet and better on the control diet. These vignettes illustrate how important a control condition is, for without it one might conclude that the specific features of the KP diet had caused improvement rather than some general or nonspecific factors associated with being on any diet. These informal summary comments aside, however, when using the ten-item scale the teachers report fewer problems for children on the KP diet than for those same children while on the control diet or during the baseline period. The fact that the informal comments sometimes convey a different picture than the scale is suggestive that the effects are not dramatic or clear cut.

Dietary Habits and Nutrient Intake

The diet diaries revealed that all the children were receiving adequate supplies of vitamins and nutrients before and during the study. Several of the children had peculiar food habits which could have led to problems over a long period, but all children received at least 66% of the recommended daily allowances of foodstuffs. The only statistically significant difference between intake on the two diets was less carbohydrate intake while the children were on the KP diet, but lack of correlation between degree of improvement and carbohydrate intake rules out decrease of carbohydrates as a cause of improvement. The amount of vitamin C consumed by children at breakfast was severely reduced during the KP diet, an effect of eliminating many of the vitamin C-containing fruits. Table II presents the results of these analyses.

Table II. Percentage RDA and Standard Deviation of Group's Intake during Pretreatment-Baseline, Control and KP Diets

	Pretreatment–baseline		Control		KP	
	Percentage RDA	SD	Percentage RDA	SD	Percentage RDA	SDA
Calories	80.4	24.8	82.1	29.1	76.6	30.4
Protein	201.7	62.3	206.3	64.2	194.3	58.5
Fat (gm)	79.5	27.4	75.5	33.7	77.6	33.5
CHO (gm)	228.5	68.9	242.6	71.5	215.3	80.5
Calcium	127.9	59.1	120.9	60.7	105.7	46.9
Iron	99.3	35.3	99.6	36.6	92.9	28.6
Vitamin A	128.5	137.7	123.7	85.7	107.3	122.8
Thiamin	95.8	43.5	91.9	47.6	83.4	36.4
Riboflavin	157.5	65.0	151.1	67.8	133.8	48.8
Niacin	84.2	38.6	78.3	35.7	94.3	48.2
Vitamin C	221.2	182.4	239.5	204.9	140.9	110.4

Follow-Up of the Children after Two Years

One of the most important methods of evaluating a new therapy is the observation of the extent to which the results are sustained over time. We encouraged parents to follow either diet, and, two years later, we contacted them by telephone. We read the items of the APQ over the telephone, and the parents were requested to score the child's present behavior. One mother had moved and could not be contacted. Only two of the children continued to follow the KP diet strictly. Both of these had shown improvement during the study. Two children had been on totally unrestricted diets in the interim. The remaining ten mothers reported that they attempted to avoid artificial colors and flavors, but not in a strict manner.

When asked about specific foods which appeared to cause behavioral changes in their children, one mother reported that natural almond flavoring, artificial vanilla (vanillin), egg noodles, and margarine all produced increased hyperactive behavior in her child. Another reported a dramatic change in her child, consisting of insomnia and nonstop talking following the ingestion of margarine or "any food containing artificial colors or flavors." This child was 14 at the time of the interview. These two mothers were the ones who were still maintaining their children on the KP diet. A third mother reported that chicken noodle or chicken rice soup, tomato sauce, Jell-o, and Pepsi-Cola caused a hyperactive reaction in her child. One mother reported that her child experienced prolonged crying spells following the ingestion of chocolate candy, cola drinks, or orange soda. Six mothers reported that sweet foods in large amounts were followed by an increase in hyperactivity; M&M's, artificial lemonade, and chocolate candy were singled out by the three mothers respectively as apparent offenders.

Thirteen out of fourteen mothers described their children's behavior as having improved since the end of the study. Typically, they reported a steady, gradual increase in self-control and a marked improvement in schoolwork. Scores on the APQ were compared with the means taken previously during the baseline period of the original study (Table III). There is a general trend toward decreased hyperactivity compared with the baseline period, but when compared with the scores from the KP diet period, the follow-up scores show a slight increase in hyperactivity. Five children had been placed on stimulant medication after the study termination, and several others received other forms of therapy.

There was universal agreement among the mothers that the strict observation of the diet was a difficult and time-consuming task. It is not surprising that only 2 out of 14 were still using this diet, but the faith of these two was unshakable. The mothers had invested a large amount of

Table III. Summary of Follow-Up Status of Patients

Subject no.	Present age	Adherence to KP diet	Behavior change after eating specific foods	APQ scores Baseline mean	APQ scores KP diet mean	APQ scores Follow-up	Medication Prior to study	Medication After study	Other treatment or evaluation
1	10	complete	natural almond flavoring, vanillin, margarine, egg noodles	18.75	6.5	11	diphenhydramine	none	none
2	14	complete	margarine, artificial colors or flavors	12.25	7.25	16	dextroamphetamine	none	none
3	9	partial	chocolate candy, cola, orange soda	12.25	5.25	14	none	none	neurological and psychological evaluation
4	9	not at all	none	15.50	10.75	10	none	none	none
5	8	not at all	none	15.50	22.75	7	methylphenidate	methylphenidate (currently)	individual and group therapy
6	12	partial	artificial lemonade	18.50	20.50	17	none	methylphenidate (not currently)	family therapy
7	9	partial	chicken rice/noodle soup, tomato sauce, Jell-o, Pepsi-Cola	27.5	4.5	9	hydroxyzine	none	none
8	9	partial	none	15.25	12.50	17	methylphenidate	methylphenidate (currently)	none
9	9	partial	chocolate candy in large quantities	16.50	19.50	12	methylphenidate	methylphenidate (currently)	none
10	11	partial	candy	25.50	19.50	23	none	none	none
11	12	partial	sweets	20.00	16.75	15	none	none	none
12	10	partial	none	18.25	18.50	14	none	none	psychological testing for learning disability and IQ
13	11	partial	sweets	21.25	12.00	23	methylphenidate & thioridazine	methylphenidate & thioridazine (not currently)	group therapy
14	9	partial	M&Ms in large quantities	10.50	8.25	15	none	none	none
				$M = 17.70$	$M = 13.18$	$M = 14.50$			

time and effort in the experiment and had received strong suggestions that there might be a relationship between diet and hyperactivity. Despite this belief in the value of the diet, there was only a slight average improvement from the pretreatment period for most of the subjects. The two children still following the diet had both been on medication prior to the study (dextroamphetamine and diphenhydramine, respectively) and had stayed off medication since the study termination. This might reflect a decrease in the need for antihyperactivity medication (about 30% of medicated children spontaneously improve during a year), or the continued benefits of the diet, or both.

Discussion and Interpretation

Of the 15 children in our initial study, two showed dramatic results during the trial. The parents of these children continued the Feingold diet over a two-year period and steadfastly maintained that violations of the diet led to immediate and noticeable deteriorations in their child's behavior pattern. One must raise alternative explanations about these findings because of some of the inherent limitations of the study. A dietary intervention is itself a dramatic event in the life of the family. Mothers could not be prevented, in our study, from knowing which diet the child was receiving, and if they held strong beliefs about the diet's value, their reports could be biased. Admittedly, the teachers were supposed to be blind to the nature of the study and its baseline—diet-crossover design. It was actually the teacher who discriminated successfully the children's behavior during the study. One might argue, however, that the mothers' enthusiasm communicated itself to the children, who in turn changed their behavior on the experimental KP diet; or that the child or parent inadvertently communicated the design to the teachers. Although unlikely in our opinion, these possibilities cannot be ruled out. The "blind" of study conditions is crucial to this issue and our study did not have a foolproof method. If the mothers were biased in favor of the KP diet, they might also inadvertently show greater enthusiasm during the interview in which global ratings were made.

Fortunately, this problem has been addressed by another study. While we were conducting our study, a team of investigators from the University of Wisconsin, lead by Dr. J. P. Harley, visited our laboratory and discussed our methods. They agreed to use the same diagnostic methods, rating scales, and experimental design in a similar double-blind crossover study. In addition, they added many refinements including an impeccable procedure for keeping the dietary conditions truly blind (Harley, Ray, Tomasi, Eichman, Matthews, Chun, Cleeland, & Traisman, 1978). They supplied *all* the food for their subjects and their entire families throughout the study, including special party treats for birthdays, school lunches, and holidays. Moreover, they supplied the food in unmarked, coded packages in such a manner that not even the study team could discriminate which was the experimental and which was the control diet. Finally, they even misled the families by including items that looked suspiciously like additive-loaded foodstuffs but which were not. In addition, they obtained a variety of objective measures of children and classmates in school and in the laboratory during the experiment.

There were many measures examined in the study, most of which showed little or no effect attributable to the special Feingold diet. However, there was a significant effect on mothers' and fathers' ratings, but not on the teachers' ratings. Of the 36 children in this study, 12 of 13 mothers and 11 of 14 fathers whose children received the Feingold diet after the control diet, rated their children as improved. Our study also showed a trend (though not statistically significant) for the Feingold-second diet order to show more improvement (Conners, Goyette, Southwick, Lees, & Andrulonis, 1976).

Even more striking in the Harley *et al.* (1978) study was the finding that all ten mothers and four of seven fathers of a younger, preschool sample rated their children as improved on the experimental diet, regardless of order. These investigators conclude cautiously that their overall results do not provide convincing support for the efficacy of the experimental (Feingold) diet. They argue that the effects are mainly apparent in the subjective parent ratings and decline sharply in teacher ratings and disappear in the more objective measures. But it seems they cannot have it both ways. If their study did indeed rigorously achieve a complete disguise of the dietary manipulations, then the parent ratings, regardless of their "subjectivity" have to be explained. The probability of obtaining such findings by chance alone is miniscule.

We have pointed out that finding that a treatment effect is more noticeable when it follows a control or placebo period, is known in drug studies with carefully concealed double-blind conditions (Conners, Eisenberg, & Barcai, 1967). Why this should be so is not clear, but it may represent the greater sensitivity of the observer after having a longer baseline for observing the deviant behavior. If the children behaved miserably while on the control diet, small changes would become more visible when they were switched to an effective treatment. The results of removing the effective treatment might be less easy to notice than instating it. This speculation needs to be tested by further experimentation.

One thing is apparent from both studies. The "dramatic" effects described by Dr. Feingold in 40–50% of children on the diet are not apparent *once a comparison is made with a control diet.* It is, in fact, true that a large percentage of children, perhaps as many as 60%, show considerable reduction in rating scores between the beginning and the end of the diet period when it is nonblind. One must remember, however, that such a drop in rating scores occurs just from readministering the ratings (Werry & Sprague, 1974). On the other hand, we did have two rather dramatic improvers who maintained their gains over a long period of time. Although far from overwhelming, these findings give us some confidence that *something* is going on which is worth pursuing. If there are *any* children whose behavior is reliably worsened by food additives, then the problem is significant. With our crude measuring instruments, we must presume that other subtle effects may be present which we have not yet detected. In the following chapter, we will study one of the improvers more carefully, and introduce a new method for asking the question of whether food additives affect child behavior.

Chapter 3

THE CHALLENGE MODEL

The Problem of Natural Salicylates

Dr. Feingold's diet (Feingold, 1975b) consisted of two classes of substances, artificial colors and "natural salicylates." This latter category was included because some patients who are intolerant of aspirin (acetylsalicylic acid) are also allergically reactive to the yellow food dye tartrazine (FD&C Yellow Dye #5). If one is constructing a diet which eliminates artificial colors then it would make sense to exclude substances which "cross-react" with food colors. Based on some German work done at the turn of the century that showed various degrees of concentrations of salicylates in fruits and vegetables, Feingold excluded a variety of these foodstuffs from his diet. There are several problems with this approach.

First, recent work appears to cast serious doubt on the methods and conclusions from which the estimates of salicylate content were derived. Studies at the University of Wisconsin (Harley, personal communication) have not verified the salicylate content of many of the supposed salicylate-containing items used by Feingold in his exclusion diet. Second, the many diverse expressions of aspirin intolerance (shock, rhinorrhea, nasal polyps, bronchial asthma, prolonged bleeding time) are not caused by compounds closely related chemically to aspirin such as sodium salicylate, propionyl salicylic acid, butryl salicylic acid, and thioaspirin (Samter, 1973). Samter points out that although a small percentage of aspirin-intolerant patients did respond to a challenge of tartrazine with intense broncho-obstructive symptoms, this reaction occurs only in those individuals with a preexisting systemic disease. This disease state is not, as was formerly believed, an *allergic* response to aspirin, i.e., it is not mediated by the substances found typically in true allergic response. Because the rationale for the salicylate part of the diet appears to have weakened, and perhaps also because this part of the diet is the most troublesome to follow, Feingold himself has tended to focus more recently on the role of artificial colors rather than on natural salicylates. With so many food items to be excluded, the probability of dietary infractions is increased. In addition, as we saw earlier, there is a risk that over a prolonged period the lowered vitamin C content of the salicylate-free diet might lead to lowered resistance or vitamin deficiency. Most investigators have chosen, therefore, to restrict their tests of Feingold's hypothesis to the role of artificial colors.

Challenging with Food Dyes

We noted earlier that there are intrinsic difficulties in evaluating the effects of an experimental diet. The procedure is cumbersome and expensive, and even if one takes the elaborate precautions used by the Wisconsin studies of Harley and co-workers, there are so many factors being manipulated that it is hard to evaluate the outcome. For these reasons a committee of experts recommended that studies "challenge" patients thought to be improved on Feingold's diet with substances presumed to create the problem (National Advisory Committee on Hyperkinesis and Food Additives, 1975). This method of placing a patient with a suspected adverse reaction to a food on a diet excluding the food, and then later challenging the patient to see if a reaction develops, is a standard method used in allergy testing. In subsequent studies, we used this challenge method rather than trying to compare different diets with each other. In this method, the children who show a clear-cut response to Feingold's diet are then maintained on the diet throughout the study and challenged at various times with foods or colors which are suspected of creating an increase in deviant behavior.

A Food Challenge Case Study

One of the two best responders from our previous study was an eight-year-old boy whom we shall call Rickie. His mother consistently noted worsened behavior while he was on the control diet and improved behavior while he was on the Feingold diet. During the original trial, his mother had commented as follows about the Feingold diet:

It has really helped him a lot. He now has no problems sleeping, no headaches like he did before. You can see a big improvement in him. He can sit and watch TV, eat meals with the rest of us, and in general do a lot of things that he was unable to do before. Even when he was on Benadryl he was unable to sleep, but now he has no trouble. As far as I'm concerned it's really great, even better than when he was on 50 milligrams of Benadryl.

Rickie's sister, age 10, and brother, age 9, had also been on the diet. The sister had been something of a behavior problem and the mother remarked that the diet "has helped her a lot, too."

During the summer before our next group study began, we had the opportunity to study Rickie more closely. We were interested to see whether we could objectively document his increase in activity level following ingestion of certain foods. To measure his activity level, we had him wear two actometers, one on each wrist. An actometer looks like a wrist watch but only registers change when the child moves. The actometers were specially devised for the study of patients with activity problems (McPartland, Foster, Matthews, Coble, & Kupfer, 1975). These instruments are calibrated precisely so that they make one count each time there are 16 movements of the limb to which they are attached.

We kept Rickie on the diet throughout the summer, and he wore the wrist actometers for eight days (Tuesday through Friday during two consecutive weeks). The actometers were constructed so that the child could not read the count, but these could be read when a small magnet was held against the back of the case. Each day his mother recorded the counts approximately 1 h after arising, for the 2-h period before lunch, 1 h after lunch, 1 h after that, and again 2 h later. She recorded the exact time she read the actometers and entered the counts on a record form. We later converted the amounts to counts per min.

On the eight days of the study, Rickie was visited by a research assistant who brought him a lunch prepared by us. On Thursdays, the lunch contained items designed to be high in artificial colors such as cherry soda pop, cheese crunchies, a peanut butter sandwich made with a glazed bun and margarine, and packaged cakes. On Tuesday, Wednesday, and Friday, the lunch contained similar items such as peanut butter but also included foods that did not contain artificially colored items, as for instance, an additive-free cookie, pretzels, or doughnuts. The assistant then stayed with Rickie during the afternoon to observe his behavior, letting him do whatever he would normally be doing. Neither Rickie nor his mother knew which days contained the offending items. Figure 9 shows the activity counts for the two challenge days and for the control days.

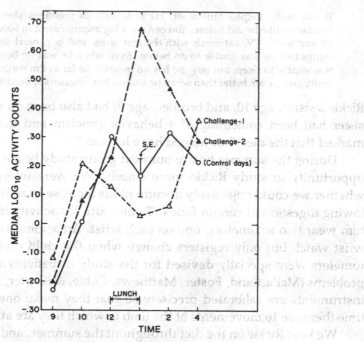

Figure 9. Wrist actometer counts on 2 challenge
and 6 control days. Scores are medians of log₁₀
scores to correct for asymmetry of score distribu-
tions. Data illustrate inconsistency of challenge response.

There appears to be a steady increase in activity during the day with a
gradual tapering off before dinner. There is a dramatic increase in his
activity after lunch on the second day of challenge, but not on the first
day.

These results illustrate one of the main problems in this research.
We frequently find strong anecdotal support for the Feingold hypothesis, but when subjected to rigorous examination the results are
not consistent. In Rickie's case, the mother's claim that even small infractions would "set him off" are verified by objective activity count on
the second day but not on the first. We would need many more such
trials before we could argue conclusively that the hypothesis was true
for this individual. Practical considerations made this impossible. In
addition, there are many problems, as we have noted, in disguising the

food contents over a long period of time. For these reasons we proceeded to a different form of challenge which is truly double-blind.

The Cookie Challenge

With the support of the Nutrition Foundation, carefully constructed batches of cookies were prepared to facilitate investigations of the role of artificial colors. The placebo cookies were matched for appearance and taste with the active cookies, which contained 13 mg of the eight colors approved for use in foods: Red #40 (38.2%), FD&C Blue #1 (3.12%), Blue #2 (1.70%), Green #3 (.13%), Red #3 (6.08%), Red #4 (.50%), Yellow #5 (26.19%), Yellow #6 (22.74%), and Orange B (.54%). The colors were supplied by H. Kohnstamm and Co., Inc., and blended with the cookie mixture by Nabisco under the supervision of one of their vice presidents. He arranged to have the cookies packaged and coded so that investigators would not know which packages contained the active or placebo cookies. He kept his code and only revealed it after the experiments were completed. Thus, neither the investigators, the children, nor the families were aware of when the children received the active challenge or the placebo.

Subjects for Trial One. Subjects for the study were solicited by a newspaper advertisement which included the ten-item checklist along with some "filler" items. Those who achieved the minimum score of 15 on the key ten items were then sent a detailed medical history. If the child's pediatrician and history revealed no contraindications (such as medications that could not be stopped), the parents were interviewed, the child was examined, and three APQ rating forms were supplied. The parents and teachers were then asked to rate the child during a one-week baseline period on Monday, Wednesday, and Friday. At the end of this period, if the child met other criteria and had baseline scores of 15 or greater on the APQ, the parents were given instructions for following the modified Feingold diet. This diet was designed to eliminate artificial colors.

We provided lists of items which were acceptable and those which had to be excluded. The dietician visited many local supermarkets and made lists of local brand names which should be avoided. When we had any doubts about particular products, the manufacturers were contacted to determine if the item contained artificial colors. When there was any

Table IV. Selection of Subjects for First Challenge Trial

Filled out initial questionnaire	117
No further response	− 44
Filled out additional screening forms	73
No further response	− 9
Began baseline phase	64
Terminated during baseline	− 1
Completed baseline/began diet	63
Terminated during diet	− 5
Completed diet	58
Terminated after diet	− 31
Began challenge	27
Terminated during challenge	− 11
Completed challenge	16

remaining doubt, we excluded the item from the diet. The modified diet is given in Appendix 3.

Subjects had to obtain a minimum score of 15 on the APQ throughout the baseline period and show a mean reduction of at least 25% after three weeks on the diet according to the parent APQ reports. Three exceptions were made: one subject showed a very strong diet effect according to the teacher but not according to the parent; two other subjects demonstrated 51% and 29% reduction of symptoms, but their baseline scores were less than 15 (13.0 and 12.7). Both children were diagnosed as markedly hyperactive by the team physician and were included despite the marginal entry scores. The children ranged in age from 55–143 months (4.5–11.9 years), with a standard deviation of 29.3 months (this is a measure of the spread around the average). In all, 15 boys and one girl ended up in the complete study, from an initial group of 166 who were contacted. Table IV shows how the children went through the selection process.

Design and Measures for Trial One. Parents filled out APQ forms on Monday, Wednesday, and Friday for the one-week baseline period. The child was placed on the diet for three weeks (diet phase) and then continued for the next eight weeks on the diet, with parents continuing to collect ratings each Monday, Wednesday, and Friday as before. During this challenge phase, the child ate two cookies per day, one at breakfast and one at lunch. Packages of cookies were supplied in weekly batches, with the batches arranged so that every two weeks the child alternated between the active and placebo cookies. A random code was used so that approximately half the subjects would receive the sequence active–placebo–active–placebo, and half would receive the opposite order (placebo–active–placebo–active).

Although we had trained pairs of observers to collect data from the child's classrooms, teacher strikes made the school data so erratic as to be relatively useless. Some teacher APQ ratings were obtained and analyzed and showed results similar to those of the parents. At one of the laboratory visits at the end of the second two-week period, and again at the end of the study, the child was tested with a device which measured his ability to keep a spot of light on a moving target while also having to respond to intermittent flashing lights on either side of the tracking task. This device, known as the ZITA (for zero input tracking apparatus) had been loaned to us by its designer, Mr. Norman Walker. It produces a tracing which reflects the accuracy of keeping the tracer on target, and a numerical score of the amount of error. A small computing device calculates the integrated area under the curve to give this score at the end of the 30-sec trial.

Results. The results for each subject on the parent ratings are shown in Table V. The table indicates that the average amounts of hyperactivity on the APQ under the active and placebo conditions were almost identical (10.15 and 10.94). Only ten teachers completed ratings,

Table V. Baseline–Diet Challenge Summary of Mean Parent Rating Scores (N = 16)

Trial 1	Baseline	Diet	Percentage improvement	Active	Placebo	Absolute difference
003	16.67	3.67	77.98	6.83	10.67	− 3.84
006	20.33	11.00	45.89	9.25	9.44	− 0.19
008	28.67	3.00	89.54	12.08	17.50	− 5.42
013	20.00	12.67	36.65	10.00	6.40	3.60
038	20.67	19.00	8.08	18.7	18.4	0.30
042	13.00	6.33	51.31	12.83	15.64	− 2.81
052	18.33	12.00	34.53	12.75	13.17	− 0.42
053	22.67	.67	97.04	5.83	11.67	− 5.84
054	17.67	6.33	64.18	8.08	6.80	1.28
061	22.00	11.67	46.95	12.00	11.75	0.25
073	15.00	11.00	26.67	15.00	12.92	2.08
085	21.33	10.00	53.12	9.33	8.83	0.50
086	16.00	2.33	85.44	8.58	9.00	− 0.42
088	12.67	9.00	28.97	5.55	4.73	0.82
090	23.67	14.00	40.85	12.30	12.50	− 0.20
100	22.67	4.00	82.36	3.33	5.67	− 2.34
M	19.46	8.54		10.15	10.94	− .79
SD	4.22	5.02		3.90	4.07	2.61

but results were similar, with active and placebo means being 11.1 and 11.4, respectively. Although there was a substantial drop in symptoms for every subject between the baseline and diet phases, with improvement ranging from more than 90% of baseline, the color-containing cookies produced no more worse behavior than the placebo cookies. Statistically, there were no effects of the active versus placebo cookies, regardless of the order in which they were received.

We noted, however, that three of the subjects had shown a marked deterioration of performance on the ZITA under the active challenge cookies when they were tested in the laboratory during the study. Figure 10 shows one of the ZITA tracings from a child under baseline, challenge, and placebo conditions. We invited these three subjects to return to the laboratory in the morning, and they were given a baseline testing with the ZITA. They were immediately given a cookie which was either active or placebo depending upon what had been received in the previous testing. If they had received an active cookie last they received a placebo, and vice versa. The testing was still double-blind because the code had not been broken. So despite the fact we knew the child received the opposite kind of cookie, we did not know which one it actually was. One hour later, the child was tested again with the ZITA. Following this trial, the child returned on a different day, was tested at baseline, given an active cookie, and tested every hour for three consecutive hours. Throughout these ZITA studies, the child had remained on the Feingold diet so that the challenges would be given against this background. The results are shown for the ZITA tests during the original trial, the day of blind testing, and the day of repeated hourly testing in Figures 11, 12, and 13.

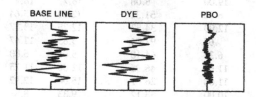

Figure 10. Visual motor tracking of child measured 1 h after eating regular diet (baseline), a cookie with 13 mg of artificial colors (dye), and a cookie with no additives. The tracings reflect attempt to track a moving spot of light.

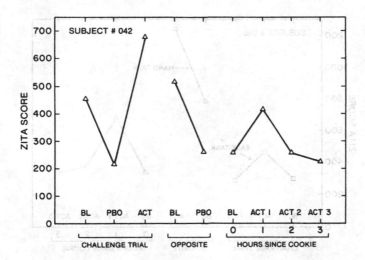

Figure 11. ZITA tracking for subject 042. Child remains on Feingold diet throughout study. Challenge with 13 mg of color in
cookies is randomly assigned. After main trial, a further comparison was made using a cookie from the opposite condition received
in the first trial. Data show deterioration under all active challenges
1 h later but not 2 or 3 h later.

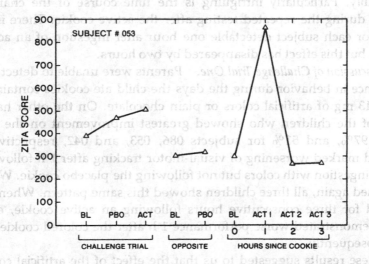

Figure 12. ZITA performance of subject 053.

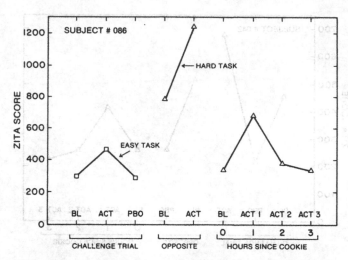

Figure 13. ZITA performance of subject 086. The child was unable to perform the difficult tracking task during the challenge trial. With practice he was able to do the hard task, but more poorly after an active challenge cookie.

The pattern of results is quite striking, for on placebo days there is either no change or an improvement due to practice as one might expect. But on trials following the active cookie, performance deteriorates noticeably. Particularly intriguing is the time course of the changes shown during the repeated testing after the active cookie. There is an effect for each subject detectable one hour after ingestion of an active cookie, but this effect has disappeared by two hours.

Discussion of Challenge Trial One. Parents were unable to detect any difference in behavior during the days the child ate cookies containing either 13 mg of artificial colors or plain chocolate. On the other hand, three of the children who showed greatest improvement on the diet (85%, 97%, and 51% for subjects 086, 053, and 042, respectively) showed marked worsening of visual-motor tracking after 1 h following cookie ingestion with colors but not following the placebo cookie. When examined again, all three children showed this same pattern. When examined for three consecutive hours following an active cookie, each child demonstrated worse performance 1 h after the colored cookie but not subsequently.

These results suggested to us that the effect of the artificial colors might be like that of a drug. Most pharmacologic agents show a dose–

time curve; that is, there is a period following ingestion of the drug when it reaches a peak and then declines over time as it is metabolized and excreted. Could the artificial colors be drugs in disguise? Perhaps only one of the colors in our "cocktail" cookies is causing the problem, or perhaps all the colors exert some mild toxic effect which produces a temporary agitated state in the children but which disappears rapidly. Such an explanation might account for the parents' inability consistently to detect worsened behavior in the children. This problem is compounded by the fact that parents and teachers only rated the children at the end of the day, and acute changes might be averaged out in their perceptions of a whole day's activity.

Adverse Reactions to the Cookies. There is another possible explanation why the parent and teacher ratings might have shown little effect due to the artificial coloring. Twelve of the 28 children who started the trial dropped out before the trial was completed. Some of these dropouts were clearly unrelated to the food coloring: one child was found to be allergic to chocolate, three children refused to eat the cookies consistently, one child became depressed and was referred for psychotherapy, and two parents failed to keep records or consistently to follow the diet. But what of the other five? If these children dropped out because they experienced severe reactions to the artificial colors, then we would have inadvertently screened out some of our most promising subjects.

Subject 014 terminated after his behavior in school went from "100s" while on the diet, to "Ds" when he began the challenge phase with the cookies. Subject 083 made a dramatic improvement in behavior during the diet phase, and as soon as he began the challenge the mother reported that he was having severe stomachaches. As a result, we reduced the number of cookies from two to one per day. He nevertheless became "wild" during the first two weeks, improved rapidly during the third week (on placebo), and worsened again during the fifth and sixth weeks (again active cookies). During the final two weeks (on placebo), he was only rated twice, but both ratings showed marked improvement. These reversals coincided perfectly with the active–placebo–active–placebo reversals. In addition to the disturbed behavior, we noted that during the office visits he had trouble breathing and had a very stuffy nose, causing him to breath through his mouth. He alternated with stomach pain, gas, and cramps during the reversals.

Subject 098 was one of the most hyperactive children in the sample. He went from the maximum score of 30 on the APQ to less than 5 during the diet phase. When he started the challenge phase (active cookies), he began experiencing numerous sleep difficulties, sneezing, and shortness of breath at bedtime. In the mornings he complained of a "thick throat" and his mother reported that he also had turned into a "wild animal." His behavior during the two-week active phase was said to be worse than at any time during the past. During the second two weeks while on the placebo cookies, his behavior calmed down. During this period, he was treated with an antibiotic (penicillin) for a sore throat, and his mother then withdrew him from the trial. The mother's ratings of his behavior are shown in Figure 14. This child had a number of the physical symptoms which Feingold has suggested are the somatic reactions which mimic allergic response, and his fluctuations in behavior seemed to alternate dramatically with the presence/absence of food coloring in the cookies.

Are these three convincing anecdotes examples of reactions to the artificial colors, or are they merely coincidences? The only way to answer this question would be a replication of the effects in these children. Unfortunately, the parents steadfastly refused to allow further investigations with their children, being fully convinced in their own minds that the food coloring was responsible for the dramatic fluctuations in symp-

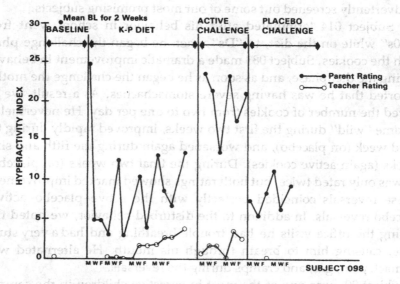

Figure 14. Behavior changes in boy who also appeared to have multiple reactions to food colors. Note absence of problems at school.

toms, and that the Feingold diet made them normal. As one parent said, "taking a miserable little kid and making him worse is no way to make friends."

Before one draws such a conclusion, however, it is instructive to look at the two other children who also developed violent reactions during the trial. Our very first subject, 001 had a literal fit of destructiveness during the first week of the cookie challenge. He cut a hole in a couch and destroyed a motorcycle with a hammer. He also broke out with severe eczema. However, this child was on the *placebo* cookies at the time.

Subject 031 also showed a dramatic improvement on the diet. Both his teacher and pediatrician commented on his markedly improved attention span, his affectionate concern for others, and his improved frustration tolerance. His mother reported a substantial improvement in his sleeping pattern during this diet phase. By Friday of the first week of the challenge with the cookies, his behavior had become "uncontrollable." His mother stopped the cookies and his behavior miraculously reverted to normal. She therefore refused to continue participation in the study. *This child also was on placebo cookies during this time.* We were so convinced ourselves that the child had shown an additive-caused reaction that we had the cookies analyzed to make sure they had not inadvertently been switched in the coding process. Our paranoia was not justified, for indeed *the cookies contained nothing but ordinary chocolate ingredients and no artificial colors.*

The lesson from these case studies is clear. Anecdotal case studies can easily be selected to bolster one's point of view. We admit that in several cases we were utterly persuaded that the food coloring produced remarkable "Jekyll and Hyde" transformations in our patients, just as Dr. Feingold had said, and just as many mothers allege after an experience with the diet; but the placebo control left us with a red face once the code was broken. It is precisely because of this capacity that human beings have for fooling themselves that science has evolved a set of rigorous safeguards in the form of experimental and statistical controls. These safeguards often seem cumbersome and unnecessary to the layman who "knows the truth when he sees it." But the lesson of the errors committed over past centuries makes most scientists unwilling to discard these rules of investigation, no matter how costly they may seem at first glance.

Summary of Trial One. After several weeks on Feingold's diet, 16 children received alternating two-week exposures to cookies that contained either artificial colors or placebo. Neither teachers nor parents could reliably detect worsening of behavior during the phases containing the artificial colors. However, a sensitive visual-motor tracking task appeared to detect reliably the food coloring in three subjects. Hourly measures of performance following a single cookie suggested that an effect was present for an hour but had disappeared by two hours. This finding suggests an acute pharmacologic type of activity which may have been missed in the global ratings by parents and teachers which were obtained at the end of the day on three days of the week. Several children showed severe reactions to the cookies containing additives, but two children also showed severe reactions while on the placebo cookies.

Chapter 4

A SECOND CHALLENGE STUDY

Challenge Trial Two

Selection of Subjects

Many parents had been screened too late to enter our first trial, but wished to begin our second trial. Others answered the newspaper advertisment and were examined in the same fashion as described earlier. Of the initial pool of 63 children for our second trial, 13 completed all phases of the study. The selection process is shown in Table VI.

Most of the children who terminated the study before the challenge phase failed to respond to the diet or showed too few symptoms to warrant inclusion. Since 37 children began the diet and 19 failed to improve,

Table VI. Summary Table for Trial 2

Filled out initial questionnaire	63
No further response	− 18
Filled out additional initial forms	45
No further response	− 4
Began baseline	41
Terminated during baseline	− 1
Completed baseline	40
Terminated after baseline	− 3
Began diet	37
Terminated during diet	− 4
Completed diet	33
Terminated after diet	− 19
Began challenge	14
Terminated during challenge	− 1
Completed challenge	13

we see that once again the figures are close to Dr. Feingold's informal estimates of 50% responders to the diet.

Again, we intended to use a cutoff score of 15 on the ten-item questionnaire and to require that subjects show a 25% reduction in symptoms during the open diet phase. There were five children whom we included despite their failure to meet these criteria: of these, one child met baseline entry criteria but had less than 25% improvement on the diet, and two were below minimum cutoffs on both measures. Sometimes a parent was astonished that our measures indicated the child was unresponsive to the diet despite the fact that they themselves had filled out the daily rating forms on which we based our conclusions. This represents a discrepancy between the parent's global estimate and daily estimates.

Sometimes the child had a marginal entry score but improved even further while on the diet. Occasionally, there were special circumstances during the diet phase which interrupted strict adherence to the diet. We decided that the five non-criteria subjects should be studied if the parents and children were motivated to follow the diet and challenge procedures. We felt we could analyze their data separately in case they showed a different pattern of response than the fully acceptable subjects. From a scientific standpoint it is better to have a sample which is clearly defined and meets specific criteria, but sometimes it is impractical to meet all criteria. As long as the sample is clearly specified, any variations in the criteria for subject selection can be taken into account.

There were four females and nine males in the second trial, ranging in age from 3.5 to 9.6 years. The mean was 74.7 months with a standard deviation of 27.9 months. Thus, this is a much younger sample than our previous trial, having a mean age of about six years compared with the eight years of the previous sample.

Design of Trial Two

The children were observed for a two-week baseline period, with parent ratings obtained each Monday, Wednesday, and Friday. (Because of the preschool age of the children and teacher strikes, we again had little teacher information available.) After the baseline period, the children were placed on the modified Feingold diet and observed for three weeks. This time the challenge was a single crossover, with six subjects receiving the placebo–active sequence, and seven receiving the active–placebo sequence. The challenge periods lasted one week each.

Because of our previous experience suggesting a possible quickly disappearing effect, we asked the mothers to give the cookies just before the child left for school, and immediately after supper. The parents were requested to observe the child closely during the three hours after supper and to fill out the rating form *at that time*. In this way we hoped to detect any transient disturbances that might be missed over a wider span of observation. Similarly, for those children attending school, we asked the teacher to pay attention to the early morning period after the child arrived.

Again the cookies were prepared and coded by an outside company which kept the code until the child had completed the trial. We taste-tested the samples to see if we could detect any differences, but our guesses were purely random, insuring us that the cookies would appear exactly the same to the child and parent at all times. Again we recorded any dietary infractions, failures to eat the cookies, illnesses, or other factors that might affect the child during the trials. Phone calls were recorded and any complaints carefully noted. Only one child terminated the study during the challenge phase of the study. She was a 4½-year-old girl who developed chicken pox.

Results

Table VII shows the average parent ratings for the different phases of the study. The scores in the two challenge conditions were corrected

Table VII. Baseline–Diet–Challenge Summary of Mean Parent Rating Scores (N = 13)

Subjects trial 2	Baseline	Diet	Percentage of improvement	Active	Placebo	Absolute difference
037	15.00	6.33	57.80	1.40	2.00	− 0.60
057	11.67	9.67	17.14	17.20	9.20	8.00
060	11.67	5.67	51.41	4.40	3.80	0.60
128	14.67	6.67	54.53	6.00	11.20	− 5.20
129	23.00	11.00	52.17	6.80	5.80	1.00
143	21.33	14.67	31.22	20.00	14.00	6.00
145	16.67	6.00	64.01	5.40	10.20	− 4.80
163	14.33	14.67	—	21.80	9.50	12.30
167	16.33	5.00	69.38	13.80	1.40	12.40
173	16.67	14.33	14.04	11.00	15.80	− 4.80
176	11.00	4.67	57.55	6.80	12.60	− 5.80
177	19.67	14.67	25.42	13.20	10.33	2.87
188	25.70	15.00	41.63	27.00	13.50	13.50
						M = 2.73
						SD = 7.11

for the level at the beginning of the challenge and analyzed by analysis of variance. The results showed a statistically reliable difference between the active challenge and placebo phases. This result was such that it would be expected by chance less than five times in a hundred. One can see from the table that the average difference between the active and placebo conditions is really rather small, only 2.7 points. But four subjects (057, 163, 167, and 184) show differences of eight or more points, which represents a meaningful difference from a clinical standpoint. The average group effect of the challenge is obviously being caused primarily by these few subjects. It is interesting to look at some of these children in more detail.

Case Studies of Double-Blind Responders

Billie (No. 167)

This boy was four years and two months of age when we first saw him. From birth he had been noted to be attentive to all stimuli and needed more than 12 h of sleep. When he was awake he was always very alert but moody and with a high activity level. He was described as an outgoing child who tended to be shy at first but warmed up rapidly. Any excitement or festivity seemed to set him off, such as parties at school or birthdays. As he got older, he became progressively more sassy and defiant of adults and began to bully other children. He was described as a "toucher" who needed to learn to keep his hands to himself.

Billie attended nursery school where his teacher described him as a very intelligent little boy who had a hard time sitting for even a minute or two. He was said to be very active and demanding of the teacher's attention at all times. He was already reading well at age four but was described by the teacher as the worst behaved of the 20 children in the class.

Billie was basically a healthy boy except for several allergies, including dust, wheat, and animals. We tested Billie with a blood test (cytotoxic test) to determine food allergies. In addition to corn, baker's yeast, and shrimp, he was also reactive to Green #3, Blue #2, and Orange B food dyes. His mother was quite sure that grape jelly was something that reliably made his behavior deteriorate.

Billie's mother made few comments during the diet phase of the study, but at one point she remarked, "At last, progress! These two days have been beautiful." This change started during the second week of the diet. She noted that though he rarely slept during the day, he had taken a nap and seemed much calmer afterwards. She made no comments during the first week of the challenge when Billie was receiving the placebo cookies, but during the second week when he received the active cookies, she commented on the second day of the week: "The effect of this cookie is lingering longer in Billie's system than I had anticipated. It isn't wearing off after a few hours as I had expected. He's waking up mornings every bit as restless, etc. as he was the evening before." The next day she commented:

> This has been a terrible week. Billie's behavior hasn't been this bad since before he started this special diet in January. Also, his appetite has decreased and he's waking up earlier in the mornings. The cookies also seem to have a cumulative effect on his behavior; e.g., Wednesday was worse than Thursday, Thursday worse than Friday. . . .

Billie stopped the cookies over the weekend and mother said;

> It appears as though having had the weekend off from eating the cookies has lessoned the degree of all the above symptoms. He started to calm down on Sunday, which indicates to me that it took about 30 hours to rid his system of the effects of the cookies. . . .

Figure 15 shows Billie's progress through the four phases of the study. It is interesting to note that the increase of symptoms during the diet phase between weeks four and five, coincided with a diet infraction (processed cheese with food coloring) and a Valentine's Day party.

Jimmie (No. 163)

This six-year-old boy became a behavior problem at about age four. He was examined at that time and found to be exceedingly distractible. For example, he would go upstairs to play and by the time he got there he would forget what he had intended to do. He was fidgety and had extreme excitement over both positive and negative situations. Jimmie was one of five children and frequently seemed to be the source of friction with his siblings. He was prone to upper respiratory infections and showed a number of immature perceptual and motor skills. The mother had five children in a six and one-half year period and seemed to be overwhelmed by the boy and his demands, along with the understand-

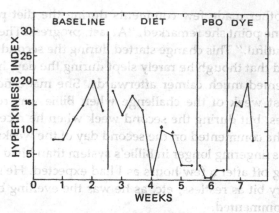

Figure 15. Four-year-old Billie's (Subject 167) parent ratings during the challenge program. Note increase in symptoms during infractions (♦) and party (▲); but mother's awareness of these events could influence ratings.

able dilemma of managing such a large, closely-spaced group of children.

Jimmie was one of the noncriteria subjects who appeared to show relatively little progress during the diet phase, but it was clear that there was a problem in supervising his diet, and there were many infractions throughout. Nevertheless, when he began the active challenge phase, his mother immediately noted on her first report that he was "compulsively loud and staccato-like, talking, skipping, and fidgeting. . . . He's had a bad period lately. . . . I'm at the point where I was going to blame it on his Wheaties." During the placebo phase, his behavior improved, showing many fewer symptoms than at any previous time. Telephone contacts indicated that Jimmie's mother was making more of an effort to control his diet because she was becoming more convinced that his behavior fluctuated with the infractions. Jimmie, like many hyperactive children, seemed to fluctuate markedly between a Jekyll and Hyde appearance, having very good days and very bad days. His mother seemed impressed that he had to exert more effort to control himself in ordinary social circumstances such as at a school play. One had the impression of a very precariously balanced control which could be easily lost. His parent rating data are shown in Figure 16.

Darlene (No. 057)

Darlene was 10½ at the time of our study. She was a slightly obese little girl who seemed suspicious and anxious during the psychiatric examination. She was slightly below the limits for hyperactivity on the APQ. She appeared to be a somewhat lonely, whiny, immature, and dependent little girl who could be oppositional with her mother, engage in minor stealing, or act impulsively, but she was not really excessively active. By the end of the diet phase, mother was checking most of her symptoms as "not at all" or "just a little." On the first day of the challenge (active cookies), the mother felt Darlene had been cooperative and that the cookies had had no effect. On the second day, she noted that behavior was "bad before and after the cookie." On the third day, Darlene's mother wrote extensive comments at the bottom of the form on which she checked most of the symptoms as "very much" present:

> Wednesday Darlene got to bed one hour late. From 5:15 to 6:00 Darlene talked over the school day—did her chores, etc. and was generally pleasant. I gave her the cookie at 6:00. After supper (6:30) I asked her to clean up the puppy mess in the cellar; she wanted the dog in the cellar. He wouldn't come. Even though I told her twice to let him upstairs she was pulling and tugging at him until he nipped her. Again I said get your job done in the

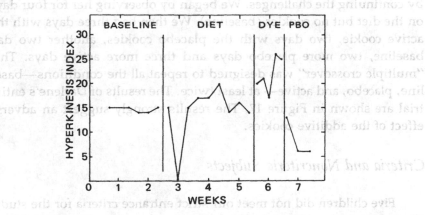

Figure 16. Six-year-old Jimmie's (Subject 163) parent ratings. Note that Jimmie has a reaction to the food dye even though he showed no response to Feingold's diet.

cellar—she finally went down, crying, kicking things, mumbling to herself. Ten minutes later I called her up to pick up her coat; she still had the job half done and was messing with something else in the cellar. The whole evening was like this—moan and groan—sulk when asked to do anything. etc.

The following day mother wrote additional comments:

This day was terrible from the morning; also it was moving day. However, Darlene had 10 hours sleep the night before—she moaned, cried, complained about any little thing all day.

She continued in this vein the next day:

Also a bad day—the cookie didn't make anything worse—no cooperation, chores half done, complains, finds all kind of excuses. She also is picking and scratching nervously.

Darlene then had a few days without cookies and then resumed taking them again, only this time the cookies were from the placebo batch. The mother's comment on the first day: "There is a complete change from last week through Tuesday!" A couple of days later she remarked how Darlene did her homework, practiced her flute "almost on her own." The next day was marked by "a good evening and cooperativeness."

Because Darlene appeared to show such clear and meaningful changes from an irritable, whiny child to a cooperative and happy disposition, we persuaded her mother to allow us to study her further by continuing the challenges. We began by observing her for four days on the diet but no cookies (baseline). We then had three days with the active cookie, two days with the placebo cookies, another two day baseline, two more placebo days and three more active days. This "multiple crossover" was designed to repeat all the conditions—baseline, placebo, and active— at least twice. The results of Darlene's entire trial are shown in Figure 17. The results strongly suggest an adverse effect of the additive cookies.

Criteria and Noncriteria Subjects

Five children did not meet our strict entrance criteria for the study. The amount of work involved in screening large numbers of children, only to end up with a small number (less than 20%) of those screened, led us to include the noncriteria subjects in the trial. As we have seen, two of those subjects (057 and 163) were among the most responsive to the challenge cookies despite being below the APQ criterion of 15 and

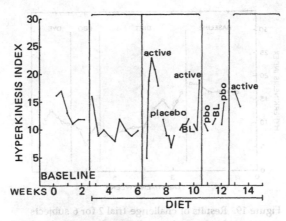

Figure 17. Darlene's (Subject 057) response to repeated challenge.

the improvement criterion of 25%. Separate analyses of variance on the criteria and noncriteria subjects showed that both groups were showing a significant effect of the active versus the challenge cookies. Figures 18, 19, and 20 show the results for all 13 subjects and for the effects of each order in which the challenge was received.

The small but reliable effect of the food dyes compared with the placebo appears in both sequences.

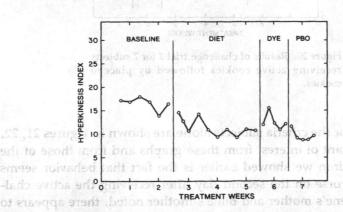

Figure 18.   Results of second challenge trial for all 13 subjects, regardless of order in which challenge was received.

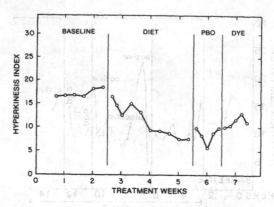

Figure 19. Results of challenge trial 2 for 6 subjects receiving placebo cookies followed by active (dye) cookies.

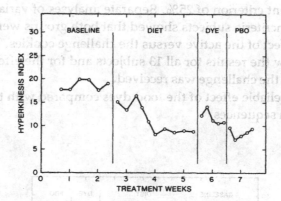

Figure 20. Results of challenge trial 2 for 7 subjects receiving active cookies followed by placebo cookies.

The data for the criteria subjects alone are shown in Figures 21, 22, and 23. One point of interest from these graphs and from those of the individual children we showed earlier is the fact that behavior seems always to be worse on the second day after receiving the active challenge. As Darlene's mother and Billie's mother noted, there appears to be some sort of cumulative effect, as if the children were "loaded" by the first day of the challenge and showed their worsened behavior thereafter. This "loading" phenomenon is well known with pharmacologic agents and could suggest that some minimum amount of the toxic substances must be present in the body before they reach the threshold for creating an effect. If we look at all 13 subjects, 11 of the children show the same or worse behavior on the second day of active challenge whereas for these same children on placebo challenge, only 4 out of 13 show worse behavior on the second day. (Three of the children showed the same behavior on the first and second days of the active challenge.) The probability of such an effect occurring by chance is quite small.

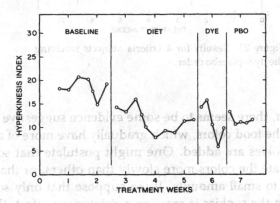

Figure 21. Challenge trial 2 result for all children meeting entrance criteria (hyperactivity score greater than 15; 25% reduction of symptoms while on open diet trial) (N = 8).

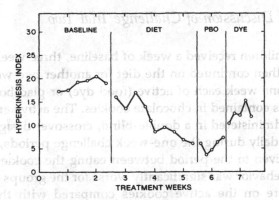

Figure 22. Results for 4 criteria subjects receiving the placebo–dye order.

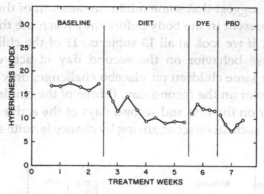

Figure 23. Results for 4 criteria subjects receiving the dye–placebo order.

Once again, there seems to be some evidence suggestive of a rather weak effect of the food colors, which gradually have more of an effect as more of the cookies are added. One might postulate that some of the children eliminate the colors more slowly than others, or that some are more sensitive to small amounts. If we suppose that only some of the food coloring in the cookies is creating a deleterious effect, then it may take larger doses before the effects become apparent immediately. All these possibilities are "after the fact" explanations and, to be convincing, we would need separate experiments to verify them.

Summary and Discussion of Challenge Trial Two

Thirteen children received a week of baseline, three weeks of Feingold diet, and then continued on the diet for another two weeks while they received one week each of active (food dye) or placebo challenge with food colors contained in chocolate cookies. The active and placebo cookies were administered in a double-blind, crossover design. Parents rated behavior daily during the one-week challenge periods, with special attention given to the period between eating the cookie at supper and bedtime. Behavior was significantly worse for the groups as a whole while they were on the active cookies compared with the placebo cookies. Four of the subjects seemed to be responsible for most of this effect. One of these subjects was subjected to another multiple crossover and found to deteriorate in behavior on the active cookies but not on baseline or placebo conditions.

Two of the best responders to the challenge were not especially hyperactive and did not show 25% reduction of symptoms while on the open diet phase. Despite this they seemed to be adversely affected by the color-containing cookies. Although the behavioral effects of the cookies are quite dramatic for a few subjects, for the group as a whole the effects are barely detectable. These findings suggest the possibility that only certain subjects are susceptible to the adverse effects of the *food dyes*, while most of the effects of Feingold's *diet* are placebo or nonspecific effects. Fifty percent of the children show improvement on the diet, but only a handful of these show the negative effects of adding food dye back into their diet.

One might argue that sustaining compliance to the diet over a long time is difficult and that noncompliance would tend to obscure the effects of the challenges. This is true, but great effort by the parents, children, and dietician kept such infractions to a minimum during the trial, certainly to a better level than would have been likely in Feingold's uncontrolled case reports. The evidence of the present trial nevertheless strongly suggests that the food dyes have an adverse effect on some of the children, above and beyond any adverse effects contributed by noncompliance to the diet. Unlike our previous trial, there were no children who dropped out of the study as an initial response to the cookies. As a result, we may have been able to include more sensitive subjects in the trial.

The sample as a whole is much younger than our first trial sample and two of the best responders were 4½ and 6 years of age. One other good responder was 10½ and she was not a classically hyperactive child. Since another study mentioned earlier (Harley *et al.*, 1978) also found that parents detected diet effects best in the younger children, this raises several possibilities for further study. The food dye effect could be a developmental effect which is only apparent at an earlier age, or the effect could be a function of body weight, with younger children essentially having smaller body mass and thus making the pharmacologic effect more potent. (Correlations of age, weight, and percent worsening of behavior between active and placebo conditions do not bear out this last speculation, although the numbers are small and the range of scores are too limited to test adequately this possibility with the present sample.) Nor is the reactivity to the food dye related to the amount of improvement in behavior on the diet. One of our best responders to the challenge (No. 163) showed no improvement on Feingold's diet but showed a severe worsening of behavior while on the active challenge. Contrarily, one of the best diet responders (No. 145) showed a 64% improvement on the diet but little response to the active challenge and a large response to the placebo.

Trial two was intended to increase the sensitivity of our observations by having parents observe the children for a specific 3-h period following ingestion of an evening cookie. On the whole, this strategy seemed to work well, since the parents did detect a poorer behavior during the evening after an active cookie compared with a placebo cookie. However, if the food dyes have an acute, quickly-decaying effect, it is hard to reconcile this with the informal observations by mothers of carry-over effects to the next day. When we looked at behavior on the second day of the active challenge, there was a significant tendency for that day to be worse than it had been on the first day—an effect quite the reverse of the one following the placebo cookies. These results raise the possibility that after the child has been on a diet which eliminates artificial colors, the first time he eats foods with such dyes in them he may not respond immediately, but have a delayed effect which depends upon a "priming" or "loading" effect from the previous day. This means that the dosage or amounts ingested, as well as the speed with which the substances are eliminated from the body, might be crucial in determining whether the child experiences an adverse reaction. Before investigating these possibilities, however, it would seem wise to replicate the present experiment. This trial showed a significant effect, but there were still small numbers of children involved, and some of these did not meet strict criteria for entry into the study. The next logical step is a repeat of the study with larger numbers.

FOODS, FOOD DYES, AND ALLERGIES

In our previous challenge with the artificially colored cookies, we found some evidence that the active cookies produced more behavior problems in the children than the placebo cookies. This effect was rather small, however, and limited to a few subjects, some of whom were not especially diet-responsive or hyperactive to begin with. Although Dr. Feingold's position has changed from his initial belief that the reaction is allergic, the small number of children actually demonstrating the effect is consistent with the idea that only a few sensitive children react adversely to the artificial food dyes. We decided, therefore, to consider the possibility that the children have some allergic response to other food substances or to the food dyes.

There have been suggestions in the literature that children with the minimal brain dysfunction syndrome (MBD) have frequent occurrences of allergy (Kittler & Baldwin, 1970). Some clinicians have argued that many of the symptoms of irritability, fatigue, and behavior problems are due to the child's allergic response to foods (Randolph, 1947). There is very little data to document these proposals. In one careful study, 79 hyperactive boys were compared with 23 normal controls with respect to general food habits, diet, and weekly consumption of 50 foods from the four food groups, including foods high in additives containing both natural and synthetic salicylates (Palmer, Rapoport, & Quinn, 1975). They found no differences in the diets between patients and controls, and there were no differences between those receiving drug therapy and those who were not. The children in the study had a 21%

incidence of allergy, which is close to the generally quoted figure of 25% for the general population.

In another study (Tryphonas & Trites, 1979), 90 hyperactive children, 22 with learning disability and eight emotional-inattentive children were tested for allergy to 43 food extracts using a special test for allergy called Radioallergosorbent (RAST) test. This test measures the amount of antibody reaction to specific allergens. This test revealed a 77% incidence of food allergy in the learning disability group which was significantly higher than the incidence of allergy in the hyperactives (47%) and emotional-inattentive group (38%). The average number of allergies in each group was 1.84, 1.77, and 0.88, respectively. There were no differences among the groups in the percentage of children who had parents and/or grandparents with a positive history of allergy, thus not confirming a genetic basis for the allergies.

One feature of this study worth noting is that as the number of food allergies for a child increased, so did the amount of hyperactivity (albeit in a rather weak fashion). This suggests that food allergies may have a cumulative effect such that an individual who is weakly sensitive to a number of foods might become hyperactive if a large quantity of sensitizing foods is eaten in a short time. This more or less random occurrence would then produce episodes seemingly unrelated to environmental events. One problem with the results of this study is the very high incidence of reactors to foods. Without a normal control group one cannot be sure that such adverse immunologic reactions are not characteristic of most normal children as well. Nevertheless, the findings are suggestive, for it is conceivable that the Feingold diet eliminates certain foods having nothing to do with additives, but which by themselves cause behavioral or physical reactions.

The Tension–Fatigue Syndrome

It has been suggested for some time that substantial numbers of people have symptoms which are directly due to food allergies or to medical problems that are aggravated by their diet (Campbell, 1973). The list of physical effects is a long one: enteritis, colitis, chronic urticaria, recurrent headache, asthma, chronic ear and sinus disease. Children with allergies are often handicapped in their studies by illness and excessive school absences. Stuffy noses, coughing, and itching interfere with their efforts to concentrate on school work. Serous otitis media, asthma, and perennial allergic rhinitis can hamper the allergic child's development as, for example, when they interfere with speech and hearing.

A number of clinicians have claimed that in addition to these physical manifestations, children also experience a "tension–fatigue" syndrome in which the child is irritable, sluggish, disinterested, and unpredictable (Baldwin, Kittler, & Ramsay, 1968; Clarke, 1948; Crooke, Harrison, Crawford, & Emerson, 1961; Hoobler, 1916; Moyer, 1975; Speer, 1958). Although the classical allergic phenomena such as hay fever and asthma may be absent, they claim that puffiness and dark circles under the eyes, generalized pallor and edematous nasal mucosa (stuffy nose) are usually present (Speer, 1963).

The allergic tension–fatigue syndrome can be partitioned into its two separate components: the allergic tension and allergic fatigue patterns. Many allergic patients complain of various types of "nervousness" or "jitters" (allergic tension). Adults speak of restlessness, apprehension, tendency to worry, feelings of inadequacy, feelings of unreality, difficulty concentrating, oversensitivity to criticism, and insomnia. Allergic children are often restless, cross and overactive. They may become timid and inclined to cry easily. Some become clumsy or uncoordinated. Allergic fatigue is often worse in the morning or after rest, but like other fatigue, may be worse after activity. Some allergic people complain of "dopiness" or torpor. Muscle aching is common. In children it usually involves the legs; in adults aching all over is common. Some patients complain of aching in the joints, especially the fingers.

The thoughtful observer will of course note that this potpourri of symptoms could arise from many different factors, and that virtually any ill known to man could be included under "tension–fatigue" syndrome. Critics (e.g., May, 1975) have pointed out the subjective reporting of symptoms, the vagueness of the clinical picture, and the lack of objective, physiological criteria for the suspected allergies.

The diagnosis of food allergy has been approached in several ways over the last several years. Usually, allergists attempt to identify allergic foods through careful history taking, patient diet diaries, elimination diets, and skin testing (Rinkel, Lee, Brown, Willoughby, & Williams, 1964; Rowe & Rowe, 1972). These methods are cumbersome, nonquantitative, expensive, and in many cases tenuously dependent upon good cooperation from the patient (Campbell, 1973). The skin (scratch) test is a procedure for attempting to detect the presence of reagenic (skin-sensitizing) antibodies, and is used by many allergists to detect specific food sensitivities. The test consists of making a tiny scratch on the skin and applying a drop of extract of the substance to be tested. A similar intradermal test consists of injecting a small amount of the extract into the upper layer of the skin. Unfortunately, such skin tests have not proven reliable (Black, 1956; Galant, Zippin, Bullock, & Crisp, 1972; Lietze, 1972; Rowe & Rowe, 1972).

However, the lack of reliable tests in the diagnosis of food allergies has not prevented allergists from using the skin tests in conjunction with diet diaries, elimination trials, and history to construct allergy-free diets. Most allergists prescribe a diet that eliminates (1) those foods suspected on the basis of the patient's history, his likes, dislikes, and suspicions; (2) those foods that are considered potent allergens; and (3) those foods indicated in positive skin tests and challenge tests (Frazier, 1974). Typically, the patient is asked to remain on the elimination diet for a period of 14–18 days, which is the time presumed necessary to clear his system entirely of allergic foods. If symptoms persist after this interval, the doctor will most likely remove still further foods from the patient's diet. Following ingestion of the test food, the patient is observed for reactions, usually with three- to five-day periods between each new food.

This procedure is fraught with problems. It requires complete cooperation by the patient and careful monitoring of the diet. What the patient does and what he reports may not be identical. The problem with children would appear to be even more complicated. It is for these reasons that attempts have been made to develop objective criteria for food allergies such as the RAST test mentioned above and a method which we decided to investigate called the cytotoxic test.

The Cytotoxic Test

During the 1950s, a test based on immunocompetence was developed by Black (1956) and subsequently refined and popularized by the Bryans (Bryan & Bryan, 1960, 1969, 1971, 1972). Their work was based on the work of previous investigators who described quantitative changes in blood cells upon contact with food antigens (Vaughan, 1934a,b, 1935, 1936). Vaughan described the decrease in the white cell count of people with severe food allergy. He formulated a test known as the "neutropenic index," but many other variables were found subsequently to affect the peripheral white count, and consistent correlations with food allergy were not found (Rowe & Rowe, 1972). Both Black and the Bryans recognized that fresh neutrophils have a normal ameboid activity under the microscope, and that when a sensitized patient's neutrophils come in contact with food antigen, a change in the morphology and activity of the neutrophils is found.

Although the laboratory reports have been available for some time, the application of the method in human clinical studies has been rare by other investigators. Recently, Ulett and Perry (1974) summarized their investigations. They reported finding that when individuals are challenged by certain foods there is a change in leucocytosis which is 50% above that of control individuals not sensitive to the foods. This effect is seen about 1½–3 h after the sensitizing foods are consumed, with re-establishment of control levels after 4–5 h. They report finding a typical dose–response relationship such that greater quantities of the offending foods produce higher rates of leucocytosis.

The reliability of this procedure has been questioned. Unless individuals can be shown to exhibit the same reactions at different times, and unless separate tests from the same blood samples give the same results, the test could not be useful clinically. In a study involving 45 patients Lieberman, Crawford, Bjelland, Connell, and Rice (1975) concluded:

> The cytotoxic food test was not found to be an accurate method for diagnosing atopic reactions to foods. Claims that the test correlated with other untoward reactions to foods (e.g., headache, diarrhea, fatigue) could not be corroborated. The test itself is time-consuming, dependent on subjective interpretation, and inconsistent in results when repetitive runs are performed on the same patient. (p. 729)

Other allergists have also doubted the test (Benson & Arkins, 1976). There are several reasons, however, that could account for the difference among the studies. There are important variations in the laboratory techinque, small numbers of cases were used to determine reliability, the selection criteria of the cases was not specified, nor the time intervals between tests and retests. Most importantly, the validity data (to determine whether the test measures what it is supposed to) are based on patient-controlled challenges that occurred at some unspecified time in the past. Basing findings on "inadvertent challenges" is a dubious method for establishing the correlation between food sensitivity and the blood reactions.

Our initial efforts with the cytotoxic test were devoted to establishing the reliability of the method, with a view to using the method for detecting sensitivity to artificial colors, foods, and additives among children. If the sensitive children could be identified objectively in advance, then the challenge of these children could specify the particular diet needed for a given child and allow much more focused investigation of the diet–behavior problem. In order to accomplish this, we first sent a technician to learn the technique from the Bryans. The method (used with certain modifications to increase reliability) is given in Appendix 4. We tested 95 adults and children with behavior problems. All testing was double-blind by the technician who did not know whether the subjects were patients or whether they had suspected allergies. They were tested on at least two occasions separated by a one- to three-month interval. Sixty-four foods and nine FD&C colors were tested. A list of foods most often found to reveal a positive reaction across patients and the correlation between the first and second testing are presented in Table VIII. Generally, behavioral scientists would accept a correlation of

Table VIII. Cytotoxic Test–Retest Results without Red Cell Reactions (RCR) Included

Item[a]	N[b]	Positive agreement[c]	Negative agreement[d]	Split[e]	Frequency (percent)[f]	Correlation[g]	Sensitivity (percent)[h]
Eggs	94	14	70	10	20	.89	74
Hops	48	1	46	1	3	—	67
Coffee	91	5	78	8	10	.79	56
Onion	58	5	52	1	9	—	91
Yellow #6	72	2	66	4	6	.83	50
Spinach	47	4	41	2	11	.96	80
Sugar beet	46	2	44	0	4	1.00	100
Banana	46	2	43	1	5	—	80
Blue #1	72	4	65	3	8	.94	73
Lettuce	55	0	51	4	4	—	—
Wheat	90	9	68	13	17	.76	58
Peppermint	48	3	45	0	6	1.00	100
Cottonseed	48	1	45	2	4	—	50
Strawberry	78	2	67	9	8	.54	31
Mustard	48	1	46	1	3	—	67
Horseradish	49	0	47	2	2	—	—
Orange	60	1	51	8	8	.50	20
Pork	94	9	75	10	15	.83	64
Carp	53	2	50	1	5	—	80
Milk	93	24	58	11	32	.92	81
Sugar cane	83	6	71	6	11	.89	67
Barley	48	0	47	1	1	—	—
Rice	48	0	48	0	0	—	—

Table VIII
(Continued)

Item[a]	N[b]	Postive agreement[c]	Negative agreement[d]	Split[e]	Frequency (percent)[f]	Correlation[g]	Sensitivity (percent)[h]
Mushroom	52	5	43	4	13	.90	71
Lobster	51	1	45	5	7	—	29
Pea	48	0	45	3	3	—	—
Peanut	59	7	49	3	14	.96	82
Red #3	56	1	54	1	3	—	67
Apple	56	1	54	1	3	—	67
Carrot	48	1	45	2	4	.82	50
Peach	48	0	46	2	2	—	—
Tuna	57	4	52	1	8	—	89
Red #4	59	2	54	3	6	.86	57
Soybean	57	0	52	5	4	—	—
Honeydew	51	1	46	4	6	.66	33
Chicken	59	3	54	2	7	.95	75
Chocolate	92	9	71	12	16	.78	60
Beef	59	2	55	2	5	—	67
Pear	50	2	47	1	5	—	80
White potato	58	2	54	2	5	.91	67
Turkey	50	0	48	2	2	—	—
Pineapple	49	1	46	2	4	.82	50
Green #3	74	1	71	2	3	—	50
Cantaloupe	46	0	46	0	0	—	—
Cherry	46	0	46	0	0	—	—
Brewers Yeast	46	0	46	0	0	—	—
Shrimp	76	7	61	8	14	.83	64
Malt	73	3	68	2	5	.95	75
Broccoli	48	0	47	1	1	—	—
Trout	49	2	45	2	6	—	67
Tobacco	53	5	44	4	13	—	71
Tomato	89	15	65	9	22	.91	77
Rye	49	0	47	2	2	—	—
Swordfish	48	0	47	1	1	—	—
Orange B	55	1	51	3	5	—	40
Bakers yeast	88	15	62	11	23	.88	73
Vanilla	47	0	46	1	1	—	—
Garlic	46	0	46	0	0	—	—
Crab	46	1	43	2	4	.81	50
Kidney bean	48	2	44	2	6	.91	50
Watermelon	48	1	45	2	4	.82	50
Lamb	48	1	46	1	3	—	67
Cucumber	50	3	46	1	7	—	86
Blue #2	72	1	69	2	3	—	50
Stringbean	49	0	47	2	2	—	—
Veal	50	4	45	1	9	—	89
Corn	92	9	72	11	16	.82	62
Radish	46	0	43	3	3	—	—

(Continued)

Table VIII
(Continued)

Item[a]	N[b]	Postive agreement[c]	Negative agreement[d]	Split[e]	Frequency (percent)[f]	Correlation[g]	Sensitivity (percent)[h]
Oat	48	2	45	1	5	—	80
Cabbage	48	1	46	1	3	—	67
Red #40	62	0	60	2	2	—	—
Yellow #5	48	0	46	2	2	—	—
Cheese	38	0	38	0	0	—	—

[a]Items: 74 items tested.
[b]N: Number of subjects with test-retest data for specified item (Red blood cell reactions were coded as missing data). Number of subjects includes hyperactive children and allergy patients.
[c]Positive agreement: Number of subjects with positive reactions to the food tested on both occasions.
[d]Negative Agreement: Number of subjects with no reaction to the food tested on either occasion.
[e]Split: Number of subjects with a positive reaction on one test and no reaction on the other: order is irrelevant.
[f]Frequency percent: Number of positive reactions relative to the total number of times the food was tested (for example, eggs: 28 positive reactions from positive agree (14 positive agree × 2 positive reactions/agreement) + 10 positive reactions from split—divided by—188 (total number of tests performed on the item).
[g]Correlation: Tetrachoric correlation.
[h]Sensitivity percent: Sensitivity of test with respect to item tested. In essence it is the probability of a positive reaction on test administration 2, given a positive reaction on test 1.

.70 or greater as satisfactory reliability, though for individual patient diagnosis a reliability of .80 or better is preferred. (By squaring the correlation coefficient, one gets the percentage agreement between the two tests; for example, the correlation of .90 represents an 81% agreement.)

These foods compare quite closely with those foods identified most often by allergists as causing allergic reactions (e.g., Frazier, 1974). The overall test–retest reliabilities of the cytotoxic test was found to be .85, based on those food items in which there were sufficient reactions to calculate a correlation. The reactions to the nine certified colors was generally found to be less than to tested foods. Only three occurred sufficiently often to calculate test–retest reliabilities: Yellow #6 (incidence 11%, correlation 0.82); Blue #1 (11%, 0.94); and Red #4 (10%, 0.86). Almost all the cytotoxic reactions to the colors occurred among the children tested.

Although we have tested over 300 adults and children with the cytotoxic test, only a small study has been completed to assess its validity. Thirteen patients with known food allergies were tested double-blind by an allergist. He used small capsules of freeze-dried foods and challenged the patients with either a food to which they were known to be allergic, on the basis of the cytotoxic test, or with a food to which they were not allergic. The patients took the food capsules double-blind over a four-week period. Ten of the patients (all adults) took the capsules according to the prescribed schedule. Nine of these ten showed a positive allergic response to the "active" but not to the control capsules, as judged by the allergist from direct physical examination. Only two of the nine reactors had an inconsistent reaction during some weeks and not others. These preliminary results give some confidence that the test may be useful in diagnosis and for constructing elimination diets for treatment. However, our first interest was to determine whether children on the Feingold diet would improve more if they were allergic to the artificial colors, and to see whether we could predict the adverse responders to the cookie challenge from the cytotoxic reactions. Feingold's hypothesis would be proved far more convincingly if one could show that children who showed evidence of being allergically sensitive to food colors were the ones who most improved on his diet and who responded with behavioral symptoms when challenged with the colors in the chocolate cookies.

A Third Challenge Trial

Selection of Subjects

Because we had no reason to believe that our responders to the cookie challenge were necessarily hyperactive (though they all had some behavior disturbance), and because some children appeared to respond to the challenge despite not responding to the diet, we relaxed these criteria for our third challenge study. We wished to include a broad array

Table IX. Summary Table for Trial 3	
Filled out initial questionnaire	92
No further response	− 21
Filled out additional initial forms	71
No further response	− 16
Began Baseline	55
Terminated during baseline	− 1
Completed baseline/began diet	54
Terminated during diet	− 3
Completed diet	51
Terminated after diet	− 13
Began challenge	38
Terminated during challenge	− 8
Completed Challenge	30

of children in order to study the relationship between cytotoxic reactions and the diet-challenge conditions. Table IX shows that we screened 92 children, with 30 completing all phases of the study. Eight children were not diagnosed as hyperkinetic: two received no diagnosis; one was diagnosed overanxious reaction; three had significant psychopathology but no specific diagnosis could be formulated; and two were judged entirely normal. These latter children were members of the Feingold Association and the parents felt they had functioned better on the diet. There were eight females and 22 males, with the mean age of 92 months (7.6 years), and a standard deviation of 27.3 months. The age range was 44–144 months.

Design and Procedure

Baseline observations were collected each Monday through Friday by both parents and teachers. Teachers rated between 9 A.M. and 12 noon, and parents between 6 P.M. and 9 P.M. The children were then placed on the modified Feingold diet for three weeks, with ratings continuing daily, Monday through Friday. The challenge occurred in one week alternating active and placebo phases, with cookies being eaten twice daily at 9 A.M. and 6 P.M. Ratings continued as before on a daily basis. Seventeen children received the placebo–active–placebo–active sequence, and 13 received the active–placebo–active–placebo sequence. Neither the parents, teachers, children, nor experimenters knew which sequence was involved until after the experiment was over. When a child had finished the trial, the executive keeping the code was called and given a guess as to which sequence we thought the child was receiving, prior to being given the actual sequence.

Results

Eight children showed less than 25% improvement on the diet, seven showed 25–50% improvement, and 15 showed 50% or more improvement. Thus, exactly 50% of the children showed a reduction of symptoms by 50% or more. Table X and Figure 24 show the results of the entire trial and challenges. It is obvious from the data that there was no effect whatsoever of the challenge, with the mean difference between the active and placebo phases being near zero.

Table X. Baseline–Diet Challenge Summary of Mean Parent Rating Scores (N = 30)

Trial 1	Baseline	Diet	Percentage improvement	Active	Placebo	Absolute difference
066	7.2	5.8	19	5.29	5.33	− 0.04
180	18.6	2.6	86	12.30	10.00	2.30
186	16.0	6.6	59	5.70	7.20	− 1.50
189	8.6	4.6	47	2.10	3.00	− 0.90
191	15.8	18.0	—	14.00	14.30	− 0.30
194	19.0	7.2	62	8.29	11.00	− 2.71
195	22.8	16.6	27	12.70	10.50	2.20
205	9.0	2.0	78	3.40	7.63	− 4.23
207	10.6	4.2	60	10.67	7.80	2.87
208	16.8	4.0	76	1.90	5.80	− 3.90
209	11.2	6.8	39	2.10	7.30	− 5.20
219	7.2	2.4	67	1.80	2.40	− 0.60
222	20.6	16.2	21	13.50	10.00	3.50
231	12.4	9.4	24	12.44	10.50	1.94
232	13.2	6.4	52	10.33	10.20	0.13
236	6.2	1.4	77	2.70	2.00	0.70
239	21.6	12.6	42	6.13	9.80	− 3.67
246	—	—	—	5.00	8.70	− 3.70
247	9.8	5.6	43	8.90	8.00	0.90
248	9.4	8.8	6	6.20	7.60	− 1.40
249	8.0	5.8	28	2.30	1.70	0.60
251	7.6	2.8	63	2.20	3.40	− 1.20
255	21.6	5.4	75	14.00	16.89	− 2.89
256	21.10	8.0	62	18.30	18.30	—
257	11.4	11.6	—	6.40	12.80	− 6.40
260	12.0	5.2	57	9.44	5.80	3.64
261	20.4	10.2	50	17.10	17.20	− 0.10
267	22.6	16.8	26	11.00	14.20	− 3.20
269	3.0	8.8	—	4.60	4.80	− 0.20
272	6.2	6.8	—	5.90	6.70	− 0.80

Reasons for Terminations

Table XI shows the various reasons why children terminated the diet. Where there was no information on terminations it was usually because we were unable to contact families that moved or simply dropped out of the study and were unable to be reached. Only subjects 170 and 242 were possible reactors to the active colors. Again, there are several children who could not tolerate the chocolate in the cookies, although we had no independent confirmation that they were truly allergic to the chocolate. On the whole, there appears to be little selection bias which might have eliminated true responders from the challenge.

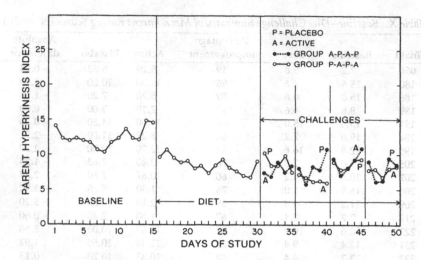

Figure 24. Results of challenge trial three. There were 17 children who received placebo–active–placebo–active sequence, and 13 who received active–placebo–active–placebo. No statistical differences occur among the treatment phases.

Effects of Challenge and Cytotoxic Reaction

Since the children had a double-crossover trial, we can consider that they actually had two trials. That is, for the first and second weeks, the child received a placebo and an active challenge (or the reverse), and for the third and fourth weeks he received another placebo–active (or reverse order) challenge. It is interesting, then, to take the average symptom scores on the APQ and to see whether they go up or down between the first and second week and between the third and fourth week. Suppose, for instance, that a child received the cookies in the placebo–active–placebo–active order. We would expect that his symptoms would first be low on placebo, increase on active, become low again, and increase again. If the child actually showed this pattern we would guess correctly that he was on the placebo–active–placebo–active sequence. For weeks one and two, if we guessed correctly, we would have a "hit"; similarly, for the third and fourth weeks. It is possible that a child would increase from the first to second week in symptoms but actually be on the active–placebo sequence, in which case we would guess incorrectly and have a "miss."

Table XI. Reasons for Termination

Terminated during diet

Subject no.	Reasons for termination:
170	Subject exhibited bad behavior on 3rd day of diet and personal family problems that made rating behavior difficult. Mother feels artificial colors are the cause of this bad behavior.
199	Subject allergic to chocolate.
210	Not able to control while subject in school.
229	Mother no longer interested in challenge because subject was not evaluated.

Terminated during challenge

Subject no.	Reasons for termination:
192	Trial 2 discontinued because of mother's illnesses.
213	Data unreliable and diet inconsistent. Subject did not eat cookies daily. Subject had difficulty staying on diet because of his age (14).
214	No information available on termination
216	Data available but unreliable.
224	Subject ill during challenge.
243	Subject ill during challenge.
259	No information available on termination.

Terminated after diet

Subject no.	Reasons for termination:
182	Mother feels that there is no noticeable difference in subject while on diet. Family is moving and will no longer have teacher ratings.
185	Mother sees no improvement while subject is on diet. Subject back on medication.
198	Subject allergic to chocolate.
217	No information available on termination.
238	Family problems (grandfather is in hospital).
242	Subject's doctor advises to terminate study because of subject's condition.
244	Mother's illness.
252	Subject has become very nervous since being on diet.
255	Diet appears to be unsuccessful in mother's opinion.
262	Subject's doctor feels diet is not helping.
263	No information available on termination.
265	No information available on termination.

In Tables XII and XIII, we have tallied the hits and misses. A plus sign (+) means that the child's response to the challenge was as expected from the kind of cookies he was eating, and a minus (−) sign means the child's symptoms went opposite to what was expected for the cookies he was eating. We have divided these into the first half of the experiment (1st trial) and second half (2nd trial). There are also two cytotoxic tests which were taken during the baseline phase of the study.

Table XII. Relationship between Improvement on Feingold Diet, Behavior Change, and Cytotoxic Reactions to Artificial Colors

Subject	1st trial	2nd trial	Percentage diet treatment	No. of color reactions	
				1st test	2nd test
180	a	b	86	0	1
186	a	b	59	0	2
189	a		47	1	2
191	b	a	0	2	1
192	c	c	3	1	1
194	a	a	62	3	4
205	a	a	78	2	0
209	a	a	39	0	c
216	c	c	100	2	3
219	a	b	67	0	1
224	c		43	0	0
231	b	b	24	2	c
232	c	c	52	2	2
236	a	c	68	1	1
239	a	a	42	1	1
246	a	a	23	2	1
247	b	b	6	1	3
248	a	c	43	2	3
249	a	b	28	0	1
251	b	c	63	3	3
255	c	b	75	1	1
256	c	a	62	1	1
257	a	c	0	0	c
261	b	c	54	0	c
267	a	c	28	4	c
269	b	c	0	c	c
272	a	c	0	c	c
Totals	7 + 14 −	6 + 8 −			

[a]Challenge guess incorrect.
[b]Challenge guess correct.
[c]Data not available.

Table XIII. Distribution of Color Reactions on Cytotoxic Test in Relation to Correct (+) and Incorrect (−) Guesses of Challenge Condition

No. color reactions	No. of subjects with hit or miss					
	Cytotoxic		Trial 1		Trial 2	
	Test 1	Test 2	No. +	No. −	No. +	No. −
0	7	3	1	5	3	2
1	8	10	2	3	2	2
2	8	3	2	4	1	3
3	2	5	1	1	0	1
4	1	1	0	1	0	0

We can see that there were seven hits and 14 misses for the first trial and six hits and eight misses for the second trial. Obviously, there is a chance relation between direction of change in symptoms and type of challenge cookie.

We can see from Table XIII that for the subjects who had two or more color reactions there are just as many hits as those with none or one reaction. If there is a relationship between the cytotoxic reactions and the food challenge, then we should get many more correct guesses and fewer misses for those children who are actually sensitive to the food coloring, as indicated by the cytotoxic test. However, this is not the case, from which we must conclude that either the cytotoxic test is invalid or that our measuring instrument is insensitive to the behavioral changes being produced by the colors. When we add the colors back into the child's diet in the active cookie phase, we would expect those with a demonstrated sensitivity to show an increase in symptoms, but since they do not, we must reject the hypothesis that the colors are causing adverse behavioral reactions.

One might argue, however, that even though the cookies do not produce significant adverse effects (as they appeared to do with the subjects of the second challenge described in Chapter 4), the improvement on the diet is substantial and that those children who show the most improvement would be those who show the most allergic sensitivity to the colors. Figure 25 shows a plot of the number of color reactions on the cytotoxic test (combining the reactions from tests one and two) against the percent of improvement on the diet. One can see that there is no relationship. If anything, the relationship is in the wrong direction, for the mean improvement for those with two or more color reactions is 43%, while the mean improvement for those with none or one reaction is 45% (which, of course, is not statistically significant).

Summary of Cytotoxic Studies

We have investigated a promising method of detecting food allergies, the cytotoxic test. We have performed this test on over 300 adults and children, many on two separate occasions. The test appears to be quite reliable. A small, double-blind, challenge study showed that in patients with known allergies, the cytotoxic test predicts to which foods they will show allergic response. Although the validity data are preliminary, they are sufficient to conclude for the moment that the cytotoxic

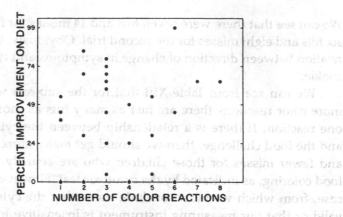

Figure 25. A plot of the number of positive
reactions to the cytotoxic test for food colors
against degree of improvement on the Feingold
diet for subjects of challenge trial 3. The plot
shows no relationship between food dye sensitivity and diet response.

test is both reliable and valid. We have tried to use the test to predict
which children will show a response to the Feingold diet, and which
children will respond adversely to the challenge cookies containing artificial colors.

Children showed significantly more cytotoxic reactions to colors
than did adults, but cytotoxic reactions are neither related to the response to Feingold's diet, nor is there a relationship to a double-blind
challenge with the artificial colors. We cannot be sure of the validity of
the cytotoxic test, but as it is our only independent measure of color
sensitivity, for the moment we must conclude that success in responding
to Feingold's diet is not related to an allergic color sensitivity. In future
studies it will be necessary to challenge subjects with individual colors
rather than in a mixture of colors as we have done. It will also be necessary to use more refined criteria of response, including somatic as well as
behavioral reactions. It is quite possible that some children will respond
to colors with some type of allergic symptom such as puffy eyes or
rhinitis, but will fail to show a behavioral reaction, and vice versa.

There is the additional possibility that response to the colors is a
graded one, depending on the amount of color ingested. Dose–response
studies need to be completed, because it is conceivable that each child

has some threshold or limit of tolerance for an amount of color, and that no response will be seen until this threshold is reached. Our studies do not vary the amount of the colors beyond those contained in the cookie formulation. We shall discuss this matter when we consider the dosage question more fully.

Chapter 6

FOOD DYES, ACTIVITY, AND LEARNING

We have not found conclusive evidence that improvements on the Feingold diet are related to an allergic response to the artificial colors, neither have we found any reliable evidence that children who eat artificial colors on the Feingold diet become worse in behavior. Although we found such an effect in our second challenge study, it was not repeated in our third challenge study. Our strongest evidence seemed to come from the sensitive measures of visual-motor tracking while the child was trying to perform under conditions of distraction. That effect seemed to be brief and transient, as if a drug had acted briefly and then diminished in activity. Since we no longer had the computerized tracking apparatus to investigate this question further, we decided to use alternative methods that would be sensitive indicators of attentiveness and activity.

Paired-Associate Learning

This test had been used by us in early drug studies and had showed an effect of stimulant drugs on the learning performance of hyperactive children (Conners & Eisenberg, 1963; Conners, Eisenberg, & Sharpe, 1964). More recently, modifications of the test have been used successfully to measure performance of children with learning and hyperactivity problems and to separate drug responders from nonresponders (Swanson & Kinsbourne, 1976; Swanson, Kinsbourne, Roberts, & Zucker, 1968). Although the test is called a learning test, in actuality it probably measures several things, including how attentive the child is, how well he remembers on a short-term basis, and how persistent he is.

In the particular version of the task used by Swanson and Kinsbourne (1976), pictures of animals are paired with a number, and the child has to learn to associate the correct picture with its number. The length of the list a child can learn is dependent on his age, intelligence, and persistence. For practical reasons, one usually uses a list that is feasible for the child to learn in a 15–20 min period. The length of the lists to be learned varies from three items for younger children to 24 items for older or more intelligent children. Each child begins with a list of 3–8 items, which is increased in length if the child can give two errorless recitations of the list after ten exposures or trials. The experimenter presents the pair to be learned (e.g., a dog together with number 4) and after several such items have been presented, the picture of the dog appears and the child must give the correct number.

The procedure is continued until the child either gets through the 24-item maximum list or fails to repeat a given list for two consecutive errorless trials. The largest list the child learns is called the "capacity list," and it is this list that is used on all subsequent sessions. During the learning process, the order of items presented in a list is randomized so that the child does not learn the sequence of items rather than just the pairings of the items. Ten seconds are allowed for a response to each picture, and if the response is incorrect the answer is provided. Scores consist of the number of trials to reach the criterion of two errorless trials and the total number of errors made during the trials.

Activity Level

We have already described the actometers used in an earlier case study. In this study, an actometer was attached to the nondominant ankle and wrist of the child while he was engaged in the paired-associate learning task. We also recorded the movements of the child as he sat in the chair while learning. This was accomplished by having a swivel chair that closed switches as the child rotated, tilted, or moved his feet.

Behavioral Ratings

While the experimenter is working with the child during the learning tasks, the child displays a variety of responses ranging from impulsive, hasty responding to dejected and distractible responding. We used a number of scales for rating these behaviors during testing (Appendix 5). These six scales are on a 1–5 basis, with total scores ranging from 5 to 30 points.

Subjects

Nine children who had participated in the previous trials were selected for further study. There were 4 girls and 5 boys ranging in age from 5–0 to 10 years. During the previous studies, eight of these children showed a clear behavioral improvement during the nonblind dietary phase of the trials compared with baseline as rated by both parents and teachers on the APQ. They also showed a differential behavioral response during the challenge and placebo phases of the double-blind trials. One child had not participated previously but was included because the parents reported a marked sensitivity to artificial colors, and because he had shown a marked improvement on the Feingold diet carried out by the parents on their own. All the subjects were tested during the summer and were following the Feingold diet during the study.

Design and Procedure

Two sessions were scheduled for the same time of the day at one- to two-week intervals. Each session began with a baseline testing period during which the child completed his first learning task of the day. Activity measures and ratings were recorded for the period during which the child was performing the learning task. Two chocolate cookies were then distributed under double-blind conditions. At a given session, each of the cookies contained either 13 mg of artificial colors or placebo. Further learning tests were then conducted at 45, 90, 135, and 180 min after the two cookies were eaten, with activity and rating measures completed each time. Four of the subjects were randomly assigned to receive the active cookies during the first session, followed by placebo in the second testing week (active–placebo order); and five received the placebo–active order.

Results

There was remarkable consistency in the ratings and two measures of activity during the course of the day. As may be seen in Figures 26, 27, and 28, activity tended to increase during the day and to level off or return to baseline later in the day. Since the children were performing a test which became increasingly more boring as the day wore on, it is not surprising that they should become more wiggly, frustratable, and active as time wore on. (This is no doubt the normal consequence of sitting in a classroom all day.) There are some differences between the first and

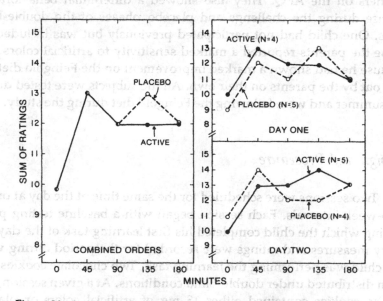

Figure 26. Activity rating scores based on child's behavior during the paired-associate learning task. The activity scale is reproduced in Appendix 5. There are no significant effects of the active versus placebo cookies which were given at time zero.

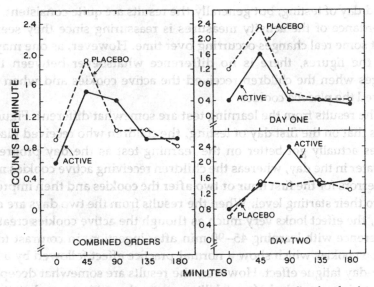

Figure 27. Actometer counts obtained from nondominant hand and wrist during paired-associate learning task.

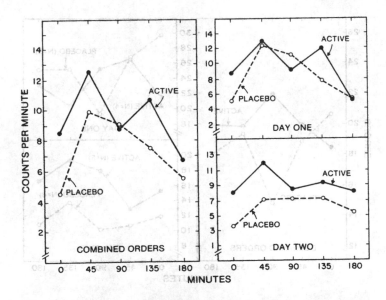

Figure 28. Activity counts taken from an activity chair which recorded seat rotations, tilting, and feet movements during learning task.

second day of testing, but generally the results are quite consistent. This concordance of the activity measures is reassuring since they seem to reflect some real changes occurring over time. However, as one may see from the figures, there is no difference whatsoever between these changes when the children received the active cookies and when they received the placebo cookies.

The results from the learning test are somewhat different. Figure 29 shows that on the first day of testing, the children who received placebo cookies actually got better on the learning test as the day progressed until later in the day, whereas the children receiving active cookies made more errors for the first hour or two after the cookies and then improved back to their starting level. When the results from the two days are combined, the effect looks very much as though the active cookies create an interference with learning 45–90 min after ingestion, in contrast to the placebo cookies which show a normal practice effect followed by a late-in-the-day fatigue effect. However, the results are somewhat deceptive, for there is a great deal of variability among the subjects, and statistical tests fail to confirm that the changes are reliable. (These tests evaluate

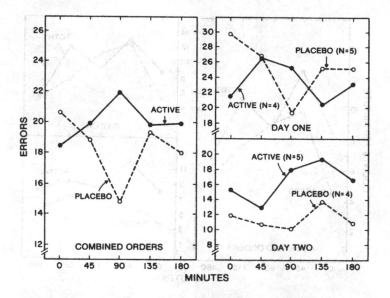

Figure 29. Error scores from paired-associate learning task. There are no significant differences between active and placebo cookies.

whether there are differences in the shape of the curves for each day separately and when combined.)

We examined each child's rate of learning as well as the number of errors they made. Although there seemed to be some children whose rate of learning slowed up after the challenge with active cookies, this was not consistent since almost an equal number of subjects showed the same effect on the placebo cookies. The experimenter who worked with the children during learning could not guess accurately which children received the active and which received the placebo cookies. On the whole, although there is some suggestion that an acute effect of the active cookies occurs within 45–90 min and then wears off, the effect is not nearly consistent enough to be more than suggestive. Perhaps with larger numbers of subjects, more tests of greater sensitivity, and so forth, we might be able to confirm the hypothesis. But these are again after the fact explanations, and as such can carry very little weight. With our 20–20 hindsight, we should have increased the amount of colors in the challenge materials. Had it been available, we also would have used the ZITA visual-motor tracking task again, preferably with the same children who showed the initial effects. But given the data we presently have available, especially since these laboratory studies were conducted on *the very best reactors* from the clinical diet and challenge trials, we must surely conclude that the artificial color hypothesis of hyperactive behavior is unproven.

Chapter 7

FOOD ADDITIVES AND BEHAVIOR

THE EVIDENCE

It would be easy for a casual reader to fasten on one or another of the several studies we have reported here and come to quite opposite opinions from those of another reader who is more impressed with a different facet of the evidence. In our first study, we compared children who were following the Feingold diet with children on a pseudodiet and found that greater improvement was seen in the children while on the Feingold diet. Score one for Dr. Feingold. A closer reading, however, shows some discomforting qualifications: the effect only occurred for the teachers, not the parents; the effect occurred more for the children when they had the Feingold diet following the control diet; parents may not have been blind to the treatment; the effect was clearly present in only 2–4 of the 15 children, depending on what level of improvement one selects. Score one for the opposition.

Many of the problems of the diet versus control design were met in our first challenge experiment in which the children remained on the Feingold diet throughout the study and received alternating two-week periods of cookies with artificial colors or placebo. In this air-tight, double-blind, double-crossover experiment, neither teachers nor parents saw any significant effects. However, the experiment turned up an unexpected finding in the best responders, whose performance worsened on a sensitive visual-motor attention task for about an hour and then improved. But how convincing is that evidence? The raw data pre-

95

sented graphically *appear* to show substantial reversal of performance in three of the children after the "loaded" cookie ingestion and not after the placebo cookies. However, these data would be more convincing if the effects could be detected in some other way, such as by direct observations of behavior or in other learning tasks. When we performed those experiments later, they failed to give any clear-cut support for an acute effect of a druglike action, even with the very best reactors of all the children we tested.

Nevertheless, when we performed a second challenge study with 13 children and looked specifically for behavioral changes during the period immediately following the cookies, parents *were* significantly able to note worsened behavior on days in which the "loaded" cookies were eaten. Some of the individual children from this trial were studied more intensively in single-case designs, and again their laboratory tracking scores showed some marked effects of active cookies compared with placebo cookies.

For those who believe that the last study firmly nailed down the evidence in favor of Dr. Feingold, our next study must have proven to be a severe disappointment. With a large and carefully screened sample, we were unable to detect the slightest difference between the placebo and active cookies in another double-blind crossover experiment. Moreover, a reasonable test of food allergy and color sensitivity showed neither a relationship to improvement on Feingold's diet nor to adverse reactions to the cookies. Thus, if one takes a conservative approach, the evidence is at best inconsistent regarding a link between behavior and food additives from our studies.

One fact has emerged consistently, however, and that is that there are indeed striking improvements reported by parents and teachers *when the Feingold diet is initiated nonblind*. In each of our studies there were at least 50% of the children who showed a clear-cut improvement after the nonblind diet phase (when no disguise attempted to hide the fact that the child was on the diet). In this respect, Dr. Feingold's assertions have proven to be correct. But the essential question is *why* there is improvement on the diet under non-blind conditions. We have been unable to prove that there is anything intrinsic to the diet such as artificial colors which accounts for the improvement. That is, the artificial colors and flavors, which are supposed to be the problems causing behavioral reactions, have not been shown to be related to improvement or worsening on the diet, at least in the dosages and schedules given. Are there any other plausible explanations why so many children should appear to improve on the diet and not worsen on the challenge?

One way to answer this question is to find characteristics associated with improvers and nonimprovers. If there were particular medical or social characteristics associated with one group or the other, it might shed some light on why some children show dramatic diet-related changes and others do not. We had obtained extensive medical, social, and demographic information on our children and their families as part of their clinical work-up. We were able, therefore, to relate these items to the baseline, diet, and improvement scores on the hyperkinesis index. There is an inherent limitation with this type of *post facto* data exploration. Since there are a large number of items to be inspected for relationships, a certain number of such relationships might turn up by chance alone. One cannot take such findings as more than hypotheses for future studies unless the relationships are so striking as to be very unlikely to be based on chance; or unless an overall test of significance takes the number of tests into account.

Table XIV presents the number of parental home experience problems as related to hyperkinesis scores during baseline or on the diet, or

Table XIV. *Number of Home Experience Problems and Response to Feingold Diet or Food Dye Challenge*

		Baseline versus diet		
			APQ scores	
No. of home experience problems	No. of subjects	Baseline (BL)	Diet (D)	Improvement (BL)-(D)
0	68	15.1	9.0	6.1[b]
1	30	14.2	8.9	5.4
2	18	14.9	12.3	2.6
3	15	18.3	15.2[a]	3.1
4	8	13.1	11.4	1.8

		Challenge versus placebo		
			Challenge data[c]	
No. of home experience problems	No. of subjects	Active	Placebo	Difference
0	31	9.8	9.9	− 0.1
1	16	8.6	8.8	− 0.2
2	6	8.5	9.0	− 0.6
3	5	7.3	7.5	− 0.2
4	1	17.2	9.2	8.0

[a]Versus 0 and 1 problems ($p < 0.004$).
[b]0 versus 2 and 4 problems ($p < 0.04$).
[c]No differences in experience problems for active, challenge or active minus challenge scores.

to improvement scores. The *home experience* score is made up of the sum of positive answers to the following questions: (1) Did either or both of the child's parents die before he was 16 years old? (2) Did the child's parents ever separate (legally or by agreement) for a period of three or more months before the child was 16 years old? (3) Did the child's parents have a divorce before the child was 16 years old? (4) Did the child ever spend three or more months living in a foster home with stepparents or in a children's home because his own parents were unable to keep him? (5) Did either of the parents ever have to spend three or more months away from home prior the child's 16th birthday because of a medical or psychiatric problem? (6) Did the child lose a close relative or friend through death in the past year? (7) Was either parent a heavy drinker (or alcoholic)?

The table indicates that there are significant differences in the hyperkinesis scores and improvement scores between those children with few (none or one) or many (three) of these negative home experiences. Poor home experience is associated with less improvement. However, there is no difference for the baseline scores prior to the diet. This would appear to suggest that having negative home experiences does not necessarily dispose the children to have more hyperkinesis (although remember that the children in the study were usually selected to have high hyperkinesis scores to begin with); but it does suggest that those with negative home experiences such as separations, deaths, or family disruption caused by alcohol, are less responsive to Feingold's diet. The second half of Table XIV shows that negative home experiences are unrelated to the effect of the challenge with artificial colors. Thus, poor home experience seems to dispose against improvement on the nonblind diet, but is unrelated to worsening with the challenge of artificial colors. It seems to us that these data strongly suggest that *the diet improvements are occurring in the children without serious home problems,* and this, in turn, would seem to suggest a placebo explanation for those who do improve on the diet. Cynically, one might say that only the spoiled brats respond to the diet, and children with serious life crises do not.

Another possibility which might explain some of our conflicting results across studies is a difference in the severity of hyperkinesis in each study. Table XV presents the hyperkinesis index scores for the diet and challenge phases for all three studies. The data show, first, that there is a systematic drop in baseline hyperkinesis scores with each trial; there is less than one chance in a thousand for the differences among studies to

Table XV. Abbreviated Parent Questionnaire (APQ) Scores for Baseline and Diet Conditions for All Three Challenge Trials

| Trial | Baseline–diet phase | | | p |
	Trial 1	Trial 2	Trial 3	
N^a	59	33	47	
Baseline	18.1	15.1	11.3	.001[b]
Diet	12.1	11.5	6.9	.001[c]
Improvement	6.0	3.6	4.4	NS
	Challenge phase			
(N)	(16)	(13)	(30)	
Active challenge	10.2	11.9	7.9	.08
Placebo challenge	11.0	9.2	8.7	NS
Difference	− 0.8	2.7	− 0.8	.025[d]

[a]Number of cases.
[b]All trials significantly different.
[c]Trial 3 different from 1 and 2.
[d]Trial 2 different from 1 and 3.

be random. Moreover, we see the reduction of symptoms while on the diet is significantly smaller for the subjects of trial 3 compared with trials 1 and 2. In other words, it appears we were successively studying less severely ill children in each of the studies. The failure of the trial 3 challenge to replicate the positive findings of the cookie challenge of trial 2 could, therefore, be due to the fact that trial 3 subjects were less severely hyperactive and less diet-responsive. The possibility that we were able to get some true responder/reactors in challenge trial 2 but not the other trials is a real one. Although plausible, this explanation is after-the-fact and must bear relatively little weight in a conservative analysis of the problem.

Almost any single study, whether positive or negative, has flaws of one sort or another. Seldom does a single study prove a hypothesis, especially where clinical trials in human subjects are involved. There are simply too many variables to control in most realistic, clinical settings. More important than any one study is the cumulative evidence from several sources. It is important, therefore, that we consider the state of the evidence from other programs, as well as those from our laboratory. We have already mentioned the various early clinical studies cited by Dr. Feingold, but those trials were not presented in sufficient detail to evaluate, and they were by their nature, preliminary and uncontrolled. Since our first report appeared, a number of other carefully executive investigations have been completed.

The Wisconsin Studies

Earlier, we described the double-blind trial by Harley *et al.* (1978), supported by the University of Wisconsin's Food Research Institute. This study is unquestionably the most ambitious and carefully executed comparison of Feingold's diet with a control diet. The project was divided into three phases: spring, summer, and fall. The ten preschool boys were followed in the summer. After a two-week baseline, children were assigned randomly either to the experimental or control diet for 3–4 weeks. Diet order was counterbalanced for each of the three samples; that is, half of the subjects got the experimental diet first, followed by the control diet, and half got the reverse sequence. Weekly behavioral ratings with our parent and teacher questionnaires were obtained. A child who was not hyperactive was also monitored in the classroom as a control to show that the experimental subjects were indeed hyperactive and to control for changing classroom activities. (However, since the observer was not blind to the identity of the experimentals and controls, this control is not really meaningful as far as insuring a true difference between a normal and hyperactive child.) Classroom observers were trained to be reliable to a high level. Direct observations as well as the teacher global ratings, were obtained.

The two diets were matched carefully for appearance, variety, nutritional value, and palatability. Compliance to the diets was increased by several stringent methods, such as removing all non-diet food from the homes, supplying all the family foodstuffs during the experiment, maintaining careful diet records, and keeping the entire family on the diet. Supplemental food for guests, holidays, parties, and even entire classrooms was provided when necessary to insure that the children were exposed as much as possible to the desired foods.

The direct classroom observations during the study showed clearly that the experimental (hyperactive) subjects were more inattentive, restless, and disruptive than their normal controls. These classroom measures showed no difference between the diets. Most of the subjects were also observed in a standardized laboratory setting, again with no diet effects. Blind observers independently confirmed that the subjects were more hyperactive than their normal classroom controls. Similarly, the battery of neuropsychological tests (memory, speed and coordination, reaction time, vigilance, attention, and basic academic skills) showed very little effect of the diets. Interestingly, however, hand steadiness and reaction time did show diet effects, although reaction time showed a different effect depending on the sequence of diets. This isolated finding cannot be taken too seriously since there were also some significant differences in favor of the control diet, differences in the opposite direction to that predicted by Feingold's theory.

Analyses revealed that there was a significant diet effect for the parent ratings, but not for the teacher ratings. However, as mentioned earlier, this effect is only significant for the subjects who got the control diet first followed by the experimental diet—a peculiar finding which we also reported in our first study. Twelve out of the 13 mothers indicated a positive response to the experimental diet in this particular diet sequence, as did 11 out of 14 fathers. Although this finding is clouded by the sequence effect, a very clear effect occurred in the preschool sample where 4 out of 7 fathers and all 10 mothers rated their children as improved on the experimental diet, regardless of the sequence.

The authors of this superb study cannot be faulted for their methodology or thoroughness. However, one might question their interpretation of results and the emphasis given to the nonsignificant findings in contrast to the significant parent ratings. It is not uncommon in drug studies to find that apparently objective measures fail to record changes readily detected by parents or teachers. Partly, this is a function of the unrepresentativeness and triviality of the so-called objective measures. Whereas parents and teachers can focus on the socially significant and meaningful sequences of behavior, integrating their observations of chains of behavior occurring over many hours, by their nature the direct observations and tests are time-limited, sample only a brief unit of behavior, and use restricted definitions which are required in order to achieve reliability. In this sense, these "objective" measures sacrifice meaningfulness and representativeness for reliability.

Nevertheless, the almost complete absence of any corroborating direct observational or laboratory data—though by no means fatal—surely diminishes one's confidence in the *dramatic* nature of the findings claimed by Feingold. Unhappily, then, the study produces a mixture of positive and negative results. It seems to introduce as many questions as it answers, still leaving in doubt the truth or falsity of the basic hypothesis. In the rules of science, an equivocal answer is usually interpreted conservatively, as the authors of this study correctly chose to do.

The hard-nosed scientist will usually accept an equivocal answer as "not proven" until further evidence is in. But the curious observer or the parent who thinks they have seen dramatic changes, will probably remain unimpressed with the negative outcomes of the study and focus on the positive results. In a subsequent investigation not yet published, Harley and Matthews (in press) completed a challenge study using nine of the best reactors from their first program. The challenge study also failed to provide any support for Feingold's hypothesis. The design, cookies, and measures were quite similar to those used by us. Neither the parent nor the laboratory tests showed any challenge effects.

The Williams Study

One of the assertions made by Feingold was that his diet was superior to stimulant medications in calming hyperactive children. A study to answer this question was designed by J. I. Williams and colleagues at the Health Care Research Unit of the University of Western Ontario in London, Ontario, Canada (Williams, Cram, Tausig, & Webster, 1978). Four treatments were administered to the 26 children of the study: stimulant medication plus active (food dye) cookies, stimulant medication plus placebo cookies, placebo pill plus placebo cookies, and placebo pill plus active cookies. Each child received each of the treatments for one week. They were first observed for one week on baseline and then placed on Feingold's diet for five weeks. The children were retested on their two best treatments for the final two weeks. Since there were 24 possible orders in which the four treatments could be given, each child was assigned to one of the possible orders (with two children acting as "back-up"). This interesting design allows one to determine the separate and combined effects of the medication and its placebo pill versus the challenge cookie and placebo. If, as is presumed, the medication causes improvement in behavior, then one would expect the medication plus placebo cookie to give the best results. Medication plus active cookie should be second best and placebo plus placebo cookie should be third best. The worst result should be placebo pill plus active cookie. This is exactly the result that was found, although only the teacher ratings reached statistical significance.

The authors concluded that stimulant medication gave decidedly better results than the diet alone. It is important, however, that the active cookies did produce significantly poorer behavior than the placebo cookies. Regardless of the fact that medication was relatively superior to the diet as a treatment, the study shows clearly the deleterious effects of the active cookies compared with the placebo cookies. This effect showed up both in the abbreviated checklist and the longer 40-item version of the questionnaire for the teachers but not the parents. The parent measures showed a trend in the predicted direction.

The FDA Contract Study

Information from the various laboratories present a perplexing array of positive, equivocal, and negative findings. The interagency committee set up to study the problem at the National Institutes of Health as well as the Federal Drug Administration (FDA), recognized that alternative approaches to the problem were desirable. The FDA awarded a contract for a new study to Dr. Bernard Weiss, a behavioral toxicologist from the University of Rochester, and two members of the School of Nutrition at the University of California at Berkeley, Drs. Sheldon Margen and J. Hicks Williams and colleagues (1978). This group had been stimulated initially by Dr. Feingold's apparent successes in the California area to investigate the problem there. They intended to use as subjects some of the many families originally studied by Dr. Feingold, but finding those subjects proved to be a difficult task. Many of the children were too old or did not meet the rigorous selection criteria.

Direct public advertisements brought in a large response, but over 75% had to be rejected—an experience consonant with our own in this regard. The project nutritionist worked with over 160 families, but eventually only 30 met all the conditions for a complete trial. Twenty-two of these children, ages 1–7, who showed marked improvement on an additive-free diet, were selected for a lengthy challenge study. Specific target symptoms which appeared to diminish on the diet, as well as positive behaviors, were selected as the measures to be followed in response to challenges.

On the basis of a local survey of foods, an estimate was made that the average daily intake of food colorings by children in the area was higher than the 26 mg proposed by the Nutrition Foundation and used in their chocolate cookies. Instead, soda-pop drinks were devised that contained 35.6 mg of 7 approved colors. Eight challenges at this dosage level were given in random fashion across the 77-day period, in a completely double-blind design.

After the baseline period of two weeks, there were eight weeks during which the children ingested the soda pop. Only on one day of each week—randomly determined—did the soda pop contain any artificial colors. There was a final week with just placebo to see if any return to baseline occurred (in case there had been cumulative effects of the challenge over the eight weeks). Home and school observations were made once a week, and daily phone calls obtained global ratings of the quality of the day as well as specific symptomatic occurrences seen by mothers. Other measures included a record of the number of times a target behavior occurred during a specific 15-min period, information about sleeping habits, physical ailments, medications, infractions of the diet, and the abbreviated parent questionnaire.

The entire study was conducted with scrupulous attention to such details as nutritional intake, compliance with the diet, foolproof disguising of the challenge vehicle, reliability of observational measures, and record keeping. The statistical analysis and interpretation of results reflects a searching, balanced, and careful weighing of all data.

The results of the study were not unequivocal: although 21 of the 22 subjects failed to support the hypothesis (that 35.6 mg of colors produces aversive behaviors in the one- to seven-year-old children), one child did support the hypothesis. That child, a 3-year-old girl, showed increases in target symptoms that were picked up on global ratings, APQ, in phone observations, and in retrospective analysis of the mother's comments. Of the six days on which the mother attempted to guess when her daughter received the drink with food colors, she guessed correctly on five. The authors note that the girl was one of the lightest and smallest children in the study, suggesting that at the dosage employed, this child might be experiencing a pharmacologic effect missed in the larger and older children. (In general, with most drugs, the effect of a given dosage is inversely proportional to the body size and mass through which the drug is distributed.) Sleep patterns were not affected in any of the children. Most of the parents were firm believers in Feingold's diet, but clearly most were disappointed by this study's outcome.

The University of Toronto Studies

Throughout the series of investigations we have been considering, we have noted that the issue of the proper effective dosage to test Feingold's hypothesis has emerged several times. Most of the studies were based on the assumption that the average child consumes approximately 26 mg/day of artificial colors, in the proportions described earlier. One of the first suggestions that this estimate is too low came from an FDA memorandum prepared by Dr. Thomas J. Sobotka in July 1976. He calculated that if one includes miscellaneous foods eaten by children, for children aged 1–5 years, the 90th percentile of intake is 121.3 mg/day. The figure for older children (6–12 years of age) is 146 mg/day. The 90th percentile is the level at which only 10% of the sample has a higher intake. It represents an estimate of what a high amount of color intake would be relative to the rest of the population. Maximum levels could be as high as 315 mg/day, and average levels at 75 mg/day.

This new information has an important bearing on all the studies reported above. If the amount of food color eaten by children is on the average much higher than originally estimated, then it would explain why positive results appear in certain studies, only to disappear in others; or why children show the effect in one phase of a study and not another (depending on what else they may have eaten or how long the materials are stored in body tissues). Drs. James Swanson and Marcel Kinsbourne at The Hospital for Sick Children in Toronto, Canada, designed a study that had several new features (1979). First, their patients were admitted to the hospital and observed while inpatients, thus insuring perfect dietary control. Second, the challenge consisted of capsules containing 100 mg of color instead of 13 mg. Third, they administered the challenge between the 4th and 6th day after the children were placed on the color-free diet.

This latter procedure requires some explanation. Many allergists work under the assumption that the optimal time to challenge with a suspected substance is 4–6 days after its elimination from the diet. The Toronto group recognized that none of the other studies had taken into account the timing of the challenge relative to the initiation of the diet.

(In our studies the children usually were placed on the diets at least 3–4 weeks prior to any challenge.) Swanson and Kinsbourne decided to test the children as if they were studying a food allergy and to test during a period when allergies to foods are often most easily pinpointed.

After a few days of familiarization in the hospital, the 20 children began the special diet. Four to six days later, they were given either additive or placebo challenges under double-blind conditions. A special rote-learning test with alternative forms (allowing it to be repeated many times) was administered one half hour before the challenge (9:30 A.M.), a half hour after the challenge, and again at 11:30 A.M. and 1:30 P.M. In the afternoon, a second challenge was given, using the opposite challenge from the morning: if the child received placebo in the morning he got active in the afternoon, and vice versa. The results showed a gradual worsening of performance after the active capsules were ingested. By 3½ h after the challenge with the active capsules, performance had dropped off by about 34% compared to the placebo challenge. The impact of the artificial colors was the same in children who were responders to stimulant drugs and in those who were nonresponders. Later studies using 150 mg of colors showed about the same interfering effect on learning as did 100 mg.

Perhaps the most important development from this group's investigations is an extension of the studies to the effects of the food dyes on brain chemistry. Using techniques employed to study the effects of drugs such as amphetamine on the brain's neurotransmitters, Swanson and Logan (1979) presented data showing that Red Dye #3 inhibits the uptake in nerve cells of all the neurotransmitters and their precursors that they tested. These chemical substances are intimately involved in the transmission of neural impulses in the brain and obviously could have important effects on behavior. The effect of the food dyes was nonspecific in the sense that all the neurotransmitters were affected but only by the red dye and none of the other dyes.

It is too early to evaluate these findings, but if substantiated as genuine neuropharmacologic effects, the implications for the study of food dye and behavioral changes are obvious. Recall that Red Dye #3 constituted only 6.08% of the formula used in making up the chocolate challenge cookies in most of the studies. It seems quite possible that the amount ingested by most children in the challenge studies could be quite low compared with the amount needed to elicit an effect. Not only is the estimate of total food color consumption in children somewhat of an arbitrary guess, but the actual amount of any specific color is probably even less accurate. Finding the dramatic impact of the red dye in animal studies narrows the field for investigation in a very helpful way.

In retrospect, dosage studies in which systematic variations of each of the separate colors were evaluated, would have been more revealing than the "shotgun" approach taken in most approaches to the problem, but as usual, hindsight is more accurate than foresight.

Summary and Conclusions

On the basis of all the evidence available at this time, in answer to the question, "Is there anything to Dr. Feingold's hypothesis?" one might answer, "Yes, something—but not much and not consistently." As far as Dr. Feingold's *diet* is concerned, the facts show repeatedly that parents and teachers rate the children as improved in behavior after the children are placed on the diet. Just as Feingold claimed, at least 50% show such improvement, but these changes appear to be due largely to placebo phenomena or other nonspecific factors. On the other hand, several studies on single cases and groups of children have shown that artificial colors produce interference with learning and performance. This latter finding may be dose-related and depend on the total amount ingested at one time, and possibly on the timing with respect to previous exposure to colors. No studies have demonstrated that this deleterious challenge effect of the artificial colors is in any way dependent on or related to the removal of the colors via Feingold's diet. Most of the studies find a small number of children among those who showed improvement on Feingold's diet also react adversely when given a challenge of the colors in double-blind fashion. These findings suggest that a rather small number of children—perhaps less than 5% of these who are genuinely hyperactive—have some specific sensitivity to the artificial colors. The "dramatic" nature of the effects has been grossly overstated by Dr. Feingold, except insofar as placebo effects are dramatic among people who are at their wit's ends with difficult and unmanageable children. Table XVI summarizes most available studies, from which one can see the degree of inconsistency in results.

The numbers of the children who are affected by artificial colors in the food supply are probably much smaller than was claimed originally. However, these figures are not very reassuring to any particular mother

Table XVI. Summary of Clinical Trials on Feingold's Hypothesis

Author	Subjects	Description	Results
Harley et al., 1978	36 hyperactive school-aged boys, 10 preschoolers	Counterbalanced cross-over comparing KP and control diet	KP diet superior only in parent ratings; order effect. Significant parent ratings for preschoolers. No objective laboratory effects.
Williams et al., 1978	24 males, 2 females; 7 probably not hyperactive	Random assignment to 24 possible orders of drug/placebo, active/inert cookie challenge combinations.	Significant drug and diet effects for teacher ratings, borderline effects for parent ratings.
Goyette et al., 1978	15 males, 1 female, hyperactives with some not meeting strict criteria	Double-blind challenge/placebo crossover study over 4 two-week periods.	No teacher or parent rating effects. Hint of effects on visual-motor tracking task; later confirmed with multiple challenge/placebo replication
Conners et al., 1977	9 males, 4 females, variable criteria	Double blind challenge/placebo crossover with one-week periods and close parent monitoring after cookie ingestion.	Significant effect on parent ratings.
Conners et al., in press	22 males, 8 females, 7.6 years mean age	Double-blind, double crossover in 1-week periods with challenge by active or inert cookies.	No parent or teacher rating effects. No relationship of diet or challenge response to food or color sensitivity as measured by cytotoxic test.
Harley & Matthews, in press	9 males who were best responders from earlier study	Double-blind crossover challenge with active or placebo cookies.	No effects on any measures.
Swanson & Kinsbourne, 1979	20 hyperactive males	Double-blind challenge with 100 mg artificial color or placebo during inpatient study with total metabolic control.	Significant impairment of paired-associate learning; no relation to previous drug responsiveness.

Weiss et al., 1978	22 1–7 year olds reporting improvement on Feingold diet.	8 challenges with 35.6 mg color, double blind, during 77-day period; target symptoms, phone ratings, direct observation, behavior counts obtained from mothers	Consistent challenge results in 1 child only.
Conners et al., 1978	9 of best responders from previous trials	Paired-associate learning, activity level and behavior ratings during sessions hourly after eating cookies with active or inert ingredients.	Some suggestion of dose–time effect on first day of testing, but nonsignificant differences.
Spring et al., unpublished	6 hyperactive boys	Subjects on Feingold diet were challenged twice for 3-day periods with 13 mg color containing cookies or placebo cookies. Ratings with APQ and global.	1 of 6 subjects became worse on the active challenge, but results could not be replicated.
Levy et al., 1977	19 male and 3 female hyperactives	Subjects tested before and after 4 weeks on KP diet, after yellow dye (tartarzine), placebo and 4 week baseline in double-blind crossover	Diet and baseline effects significant on ratings but not on objective tests. Challenge effect nonsignificant; but 13 children with 25% reduction of symptoms on the diet phase showed a significant challenge effect on parent ratings.
Levy & Hobbes, unpublished	7 males, 1 female, age of 5 years, 2 months	Challenge or placebo with cookies containing 1 mg tartrazine over 14 days (4 mg/day).	13% increase in symptoms on the active challenge, but this was not statistically significant for this sample size

who wonders what she should do about such a potential hazard with her child. Pediatricians and family physicians are placed frequently in a difficult situation by mothers demanding to have the Feingold diet rather than some other treatment being offered. Large numbers of parents have turned to "Feingold associations" and have adopted a quasi-religious belief in the efficacy of the diet for all their children's problems.

We have tried to show in this book that patient and serious attempts to find evidence in favor of Feingold's basic hypothesis have been carried out in many laboratories; that no clear-cut and consistent evidence has emerged in favor of the hypothesis; but that new knowledge has delimited the questions and focused research in ways not possible before Dr. Feingold raised his alarm. We are, therefore, left with advice for parents and professionals which is based on the state of the art as we know it in medicine and psychology. A few simple caveats would seem to be in order for those who must make decisions about their children or patients.

First, it makes sense, obviously, to see that the child has a proper diet. Any elimination diet, such as Dr. Feingold's, should be undertaken only under the supervision of a pediatrician or trained nutritionist after a thorough evaluation of the problems presented by a particular child. This generally requires a careful medical, family, and developmental history; a physical examination; an interview with the child; information from the school and its support services; and careful review of home and environmental factors that might be contributing to the problem. Many behavior problems of the children improve spontaneously, but many do not. Only a qualified professional can sort out the many factors that determine the outcome for a particular child.

Second, it does no harm to provide a diet with the minimum of artificial ingredients, as long as sound nutritional principles are followed. In the same manner, it makes sense, but may not always be possible, to reduce exposure to carbon monoxide and lead fumes from automobiles; to avoid radiation from industrial sources; to avoid foods contaminated with pesticides; to stay away from lead in paint; and to limit known hazards like cigarette smoke (either in the pregnant mother or in the child's immediate environment). These industrial hazards are the unfortunate penalties we all pay for "civilized" life, and any sane person would wish to minimize their effects on children, whether proven or merely suspected. However, avoiding proven therapies because of justifiable ecologic concerns makes no sense either.

A miniscule proportion of our nation's federal budget is allotted for research on factors affecting the growth and development of children. Foods and food additives are no exception. We all owe Dr. Feingold a debt of gratitude for focusing attention on the research needed to advance knowledge in this area and to protect the heirs to our planet. The evidence has not been favorable to his hypothesis in our opinion, but his general advocacy on behalf of children deserves to be supported by all citizens through their support of efforts to increase research knowledge in this important area.

CHILDREN'S PSYCHIATRIC RATING SCALE (WITH PERMISSION FROM NIMH)

1. Tension (do not include fidgetiness)
   Musculature appears taut, strained, or tense, fingers clothing; clenches jaws; grips arms of chair; hands tremulous
2. Underproductive speech (rate amount of speech only, not rate or relevance)
   Fails to answer questions; monosyllabic; has to be pushed to get an answer, does not elaborate, blocked
3. Fidgetiness (do not include tics)
   Wriggles, squirms, moves, or shifts restlessly in chair
4. Hyperactivity
   Has difficulty sitting in chair; gets up; moves fast, vigorously; impulsive bursts of locomotion. Exclude slow ambling even if constant. In rating degree of overactivity, consider the ease with which the hyperactivity can be controlled.
5. Hypoactivity
   Few or no spontaneous movements; sluggish; movements are slowed, feeble or labored; requires prompting for initiation of motor movements; long latencies of appropriate motor behavior
6. Distractibility
   Distracted by usually minor, irrelevant stimuli; shifts from one topic to another; interrupts thought or action abruptly
7. Abnormal object relationships

113

Autistic use of objects with disregard for usual function; stereotyped and repetitive sequences or fragments of play; aimless behavior without organizing goal idea

8. Oblivious of examiner, preoccupied; facial expression and behavior do not respond directly to examiner; attention focus is oblique and vague in direction, with avoidance of eye contact; responses are very delayed and require forceful stimuli. (The fact that the child may have peculiar interest in examiner, such as obsessive interest in parts of body or clothing does not preclude a rating of withdrawal).

9. Goes along with whatever examiner says in a passive fashion, even contradicting self; does not assert self in a reasonable manner.

10. Negative, uncooperative
   Active opposition and resistance to examiner's initiative (differs from withdrawal and oblique avoidance); guarded, evasive replies, teasing, manipulative or hostile refusal to cooperate; child may remain silent in passive-aggressive fashion

11. Angry affect
   Irritable, touchy, erupts easily (shouts angrily, screams at examiner, overtly and directly hostile)

12. Silly affect
   Clowning, inappropriately giddy, playful, silly behavior

13. Confusion
   Confused, bewildered, perplexed in behavior or verbal expression

14. Disorientation
   Child is unaware of identity of surroundings after being told where he is; not aware of time discriminations; does not know age or surname

15. Clinging behavior
   Clinging, in physical and verbal behavior with the examiner; seeks physical contact; demands constant direction

16. Unspontaneous relation to examiner
   Responds to examiner, but does not initiate social or verbal overtures, or sustain conversation once begun; lacks spontaneity; restricted

17. Suspicious affect
   Expresses concern about the intent of the examination, questions instructions and good will of interviewer.

18. Depressed demeanor

Exhibits a dejection, depression in mood; looks sad; seems to be in a state of painful dejection

19. Blunted affect

   Restricted range and intensity of emotional expressions; blank or fixed facial expression; monotonous voice

20. Lability of affect

   Can suddenly vary from calm or silly to sullen mood, to screaming, crying, loud complaining

21. Pressure of speech

   Speech is hurried, accelerated, pushed, difficult to interpret

22. Level of speech development (do not include diction, rate of speech, or relevance of speech)

   From age-appropriate (1) to severely regarded (7) speech development. Using your clinical judgment of verbal IQ, estimate the level of speech development (in percent) in relation to verbal IQ.

   1 = Over 90%   5 = 31–45%
   2 = 76 –90%   6 = 15–30%
   3 = 61–75%   7 = Less than 15%
   4 = 46–60%

23. Stuttering

24. Low voice

25. Loud voice

   Voice loud, boisterous, shouting

26. Mispronunciations

   Lisping, mispronounces letters such as r, s, l; unclear speech

27. Other speech deviance

   echolalia; questionlike melody; neologisms; sentences fragmented, unusual syntax

28. Rhythmic motions (stereotyped)

   Rocking, whirling, head banging; rolling, repetitive jumping hand movements, athetoid, twiddling; arm flapping

29. Expressed feelings of inferiority

   Describes feelings of inadequacy, inferiority; self-deprecating; self-belittling.

30. Expressed feelings of grandiosity

   Exaggerates own value, boasting; unduly pleased with own achievement; says he is much better than others; distorted sense of own capacity.

31. Physical complaints
    Somatic complaints of headaches, stomachaches, dizziness, not feeling well, etc. (do not include fatigue)
32. Obesity
    Judge from child's appearance from normal physical appearance to severe obesity
33. Other eating problems
    Picky, fussy, many dislikes, extremely restricted diet, peculiar food tastes
34. Separation anxiety
    Ease with which child separates from mother or other significant people, extent of observed or reported anxiety (by child) experienced by child when separated from mother or other significant people
35. Depression
    Admits feeling sad, lonely, feels like crying, expresses a despondent or despairing attitude; difficulty in anticipating success and enjoyment
36. Euphoria, elation
    States he feels terrific, great; elevation of mood; hypomanic state. "This is the best of all possible worlds." Feels elated and wonderful; nothing is impossible
37. Lack of energy
    States he feels sluggish, fatigued; everything is too much; weary and feels unable to make slightest effort. (Do not infer from motor retardation or expressed indifference)
38. Preoccupation with topics of anxiety
    Says he has nervous or scary feelings, concerns, apprehension, fears; says he worries about failure or other mishaps; thinks about something happening to self or parents—illness, injury, death, loss, or separation.
39. Preoccupation with depressive topics
    Preoccupied with feelings of inadequacy and inferiority; expresses feeling that nothing can turn out all right; preoccupied with feelings of uselessness, futility, and possible guilt; suicidal preoccupation
40. Suicidal attempts
    0 = Not assessed
    1 = None
    2 = Suicidal threat

3 = One minor gesture without danger

4 = A couple or several minor gestures without danger

5 = Dangerous gesture

6 = Inflicting of life threatening damage to self

7 = Several life threatening attempts

41. Fears and phobias

Irrational morbid fears of specific objects, person, or situations, which, if extreme, lead to avoidance behavior; rate 6 or 7 only when fear is so severe it leads to phobic avoidance

42. Compulsive acts

Acts or "habits" which are regarded as unreasonable by the child, such as counting, checking, rituals, excessive orderlines, and cleanliness

43. Nervous habits and mannerisms

Stereotyped movements; rituals which are not perceived as irrational; facial tics or mannerisms; biting nails, fingers, cuticles; sucking of objects or body parts (thumb, fingers, hair, etc.); picking on skin, scabs, nose, twisting hair

44. Obsessive thinking

Inability to "turn off" repetitive thought; preoccupation, ruminations about abstract problems or personal matters

45. Solitary interests

Interested in activities which require little if any peer interaction, such as stamp collecting, movie-going, reading, school work, solitary activities

46. Lack of peer interaction

Isolated from other children; has no friends or cannot name current close friends or describe participation in play with peers; lacks interest in peers

47. Gang activity

Joins in antisocial activities along with a group of children (fighting, trouble making, stealing) as a cooperating group against others

48. Fighting with peers

Says he frequently gets into fights—beats up other kids or gets beaten up; says he has a bad temper

49. Bully

Says he's always the leader, winner; says he teases, bullies children; pushes children around; threatens them

50. Temper outbursts

Admits to feeling angry, irritable, touchy; admits he has a temper

51. Scapegoat
Says he's picked on, teased, left out or pushed around, and bullied by other children; may be called "sissy" or "baby"

52. Lying
Contradicts self in ways indicative of effort to hide the truth; reports telling tall stories, fibs, or admits he's accused of telling lies

53. Exploitative relationships
interested in other people insofar as he can get something out of them; callous and calculating in interpersonal activities

54. Inability to fall asleep
Reports long time to fall asleep after going to bed.

1 = Not present   5 = 46–60 min
2 = 10–15 min   6 = 60–90 min
3 = 16–30 min   7 = Over 90 min
4 = 31–45 min

55. Other sleep difficulties
Nightmares, early morning awakening, sleep walking, interrupted sleep

56. Bedwetting
Rating is for frequency of bedwetting for past 7 nights:

1 = None   5 = 4 times
2 = One time   6 = 5 times
3 = 2 times   7 = 6–7 times
4 = 3 times

57. Ideas of reference
People are looking at him, following him, staring, etc; malevolent intent is not necessary but may occur

58. Persecutory
Feels people have it in for him, try to hurt him; in the extreme rating, thinking has a delusional quality in that belief is impervious to change, rational arguments, or corrective experiences

59. Other thinking disorders
Irrelevant speech; incoherent speech, or loose associations

60. Delusions
Delusional belief or convictions besides paranoia (58), i.e., believes has introjected persons or objects in his body; has a mission, is some other person or character; has unusual powers; is guilty of some event

61. Hallucinations

The overall rating is a frequency rating reflecting the constancy of the experience:

1 = Not present
2 = 1 time
3 = 2 times
4 = 3 times
5 = 4–5 times
6 = 5–6 times
7 = Daily recurrent phenomenon

62. Peculiar fantasies

Morbid or bizarre fantasies and preoccupations, peculiar body sensations, disturbances of body image experiences (not figure drawings); preoccupation with flying, supernatural influences, sadism, masochism.

63. Lack of insight

Is convinced of the reality of hallucinations or fantasies.

Each item is rated as: not assessed, not present, very mild, mild, moderate, moderately severe, severe, or extremely severe.

STUDY PROCEDURES AND INFORMED CONSENT FOR DIET VERSUS CONTROL COMPARISON

This appendix is included for the reader who may wish more detail regarding our study procedures and the specific diets used for comparison in the first study reported in the text.

Diet Study Flow Sheet
Revised December 4, 1974

I. Referral
   1. Basis of study explained
   2. Referral information obtained
   3. Informed that school information will be obtained—parents sent release of information forms to be signed and mailed back
   4. Informed that 24-h recall will be obtained during first appointment
   5. Appointment scheduled with coordinator in one week
   6. Packet with letter to teacher and release of information form (signed) prepared and mailed
      a. Pretreatment: Teacher Questionnaire 000, Abbreviated Teacher QS 013–043

    b. Baseline: Abbreviated Teacher QS 053–083

    c. Diet 1: Abbreviated Teacher QS 093–113, Teacher Question-
naire 123

    d. Diet 2: Abbreviated Teacher 133–153, Teacher Questionnaire
163

  7. Teacher contacted and role is reviewed

II. First Appointment

  1. Parents sign informed consent.

  2. Parent questionnaire obtained.

  3. Prior knowledge questionnaire obtained.

  4. Medical and social history obtained, IQ 85 verified.

  5. Allergy history obtained.

  6. Instructed to collect Abbreviated Parent Questionnaires on
each Tuesday and Saturday for the next four weeks. If child is
on medication, that is to be discontinued on the first day of the
third week. Parents called on the first day of the third week to
remind them and answer any questions.

  7. Second appointment scheduled for one month (during fourth
week).

  8. Letter to physician mailed.

  9. Instructed to collect diet diary on Monday, Tuesday, and Wednesday of week one, and Thursday, Friday, and Saturday of
week three. Phone call made on first day of third week.

  10. 24-h recall and food frequency measure of past month obtained.

  11. Dietary questionnaire obtained.

III. Second Appointment

  1. Physician performs psychiatric, neurologic, and physical examination.

  2. Instructed to collect abbreviated questionnaires on each Saturday
for the next three weeks, and a Parent Questionnaire on the
Saturday of the fourth week.

  3. Diet assigned and requirements explained – instructed to collect
diet diary on Monday, Tuesday, and Wednesday of week one,
and Thursday, Friday and Saturday of week three. Parent called
on the first day of the third week to remind her to collect diet
diary and answer any questions. A list of infractions is to be kept
throughout the period.

  4. Food frequency measure obtained for past month.

5. Third appointment scheduled for one month (during fourth week).
6. Postcard mailed to teacher to remind her that questionnaires need only be filled out on Fridays from now on.

IV. Third Appointment

1. Instructed to collect abbreviated questionnaires and Parent Questionnaire (see above).
2. New diet assigned and requirements explained (see above).
3. Food frequency measure obtained for past month.
4. List of infractions collected or verified.
5. Dietary degree of difficulty questionnaire obtained.
6. Fourth appointment scheduled in one month.
7. Parents meet with Principal Investigator and fill out joint CGI.

V. Fourth Appointment

1. Files checked to insure all data has been collected.
2. Food frequency measure obtained for past month; list of infractions collected.
3. Dietary degree of difficulty questionnaire obtained.
4. Parents meet with Principal Investigator and fill out joint CGI.
5. Parents are debriefed and interviewed to determine the amount of prior knowledge. Prior knowledge questionnaire obtained.
6. Parents instructed on follow-up program for their child.
7. Principal Investigator and coordinator complete discharge summary for each child.
8. Summary statement dictated and copies sent to physician, school, parent.

Informed Consent

I, the parent (or legal guardian) of _____ agree to participate in the study of dietary factors in hyperkinesis. I understand that the purpose of the study is to determine whether either of two special diets is beneficial to my child's behavior, especially in reducing his overactivity and attentional problems. (Message from the Principal Investigator: "In our experience some children improve on one diet and not the other, but a child may show no change to either diet, improve on one and not the other, or improve on both. In this study we hope to find out which children do best on which diet, and whether one diet is better in an overall sense than the other".)

I also understand that the study will last two months, with one month on each diet. I have agreed during this time to adhere as strictly as possible to the prescribed diets, with the understanding that I may consult the study coordinator or the study dietician regarding any problems or questions. It has been explained to me that both diets are nutritionally balanced and pose no medical or nutritional hazards. I may withdraw from the study at any time that I or my family physician deems unadvisable for any reason whatsoever. (Message from the Principal Investigator: "If your child has been receiving medication to help his hyperactivity, this medication will have to be stopped during participation in this study. Undoubtedly there is good reason you are treating your child with medication already, so you should be aware of the changes that might occur when medication is stopped. As you know, some children are more impulsive, restless, distractible, or troublesome without this form of treatment. You will have to expect that such behaviors may recur when medication is withdrawn. However, you will be free to reinstate the medication whenever these symptoms become too troubling to you or your child, or whenever its absence threatens his school progress of safety in any way, as determined by you, the child, or your physician. We will assist you in every way to manage these symptoms so that you may discover whether one of the diets may obtain comparable or better results.")

No representation has been made to me that these diets will cure the problems of my child, and I fully understand the experimental nature of the program. I have been told that the results of any tests, examinations of treatments will be available to me, and that the progress of the study will involve the cooperation and supervision by a qualified psychiatrist, clinical psychologist, and dietician.

I have been told that ratings of my child's school behavior and performance may be sought as part of the study, and have agreed to give permission to obtain this information for the purposes of the study. All medical, social, and scholastic information or records will be open to me upon request and will be kept strictly confidential. Such records will only be available to the members of the study team or professions or persons for whom I grant specific consent. No publication of any documents or reports will in any way identify my child or his family by name unless this is specifically authorized in writing by me.

I have not been promised any reward, inducement, or payment to participate in the study, but have been told that the examinations, tests, and interviews connected with the study are free. Any services required beyond the study itself are my responsibility, and I have been told that such services or advice will be given as far as possible within the study, but should further services be required the study team will assist in locating them.

I have been told that this study is supported by the National Institute of Education of the Department of Health, Education, and Welfare, but that the study team assumes full responsibility for the conduct and execution of the study, and the Government of the United States does not assume any responsibility for any medical or legal disputes that may arise out of the study.

Signed _____

Witnessed _____

Date _____

Diet Diary Procedures

Thank you for helping in this test of two diets. Attached is a diet diary form which must be filled out for _____ days during the _____ week of the diet trial period. Please keep the record for _____ . Everything the child eats and drinks needs to be recorded.

1. Record the *time* of day for meals and snacks.
2. Record the *place* where the food was eaten (home, school, restaurant, outside house).
3. Record *what* the food or beverage is (milk, hamburger, bun) include the brand name if known.
4. Record the *description or method of preparation* (milk—whole, hamburger—broiled with butter). If the food is a casserole or mixture, list the main ingredients (macaroni and cheese—enriched macaroni, cheddar cheese, whole milk).
5. Record the *amount* of the food eaten by weight or common household measure (3 ounces, 2/3 cup, 1 tablespoon).

At the *end of the day* or *after each meal,* record all the child has eaten, asking him about foods he ate away from home and using your combined memories for what he ate while at home and at school. Write it down by meal and snacks. Also record how well you feel he followed the diet and meal and snacks. Also record how well you feel he followed the diet and other comments (child sick, child followed diet easily).

At the end of the three days, please mail the records in the preaddressed envelope provided. Please call us if you have any questions or problems.

Instructions for Management with the N Diet (KP Diet)

1. The diet lists all foods and products that are to be prohibited.

2. Any item not on the list is permitted. It is necessary to check every food package or container very carefully for artificial colors and flavors. In many cases, specific flavor or color is not indicated—avoid foods labeled with artificial or imitation flavors and artificial or U.S. certified colors. When in doubt, do not experiment—eliminate food items.

3. With the difficulty in obtaining many food items on the market that conform to the allowed food list, it will be necessary to prepare many foods at home. Examples of this would be main dishes, candy, soups, puddings, and cakes. All beverages must be checked very carefully.

4. The greatest success is usually observed when the entire family follows the diet. This removes the temptations from the child and facilitates compliance with the diet.

5. It is important that the child and family continue maintaining adequate nutrition during the study. For this reason, the foods on the *allowed list* have been grouped according to their contribution to the diet. A balanced diet which will maintain the nutritional status of the individuals following the diet can be achieved by choosing a certain number of foods from each group, to be eaten each day. The minimum number of servings from each group is suggested as follows:

Child	*Adult*
4 cereal and grains	4 cereal and grains
2 fruits (1 good source of vitamin C)	2 fruits (1 good source of vitamin C)
2 vegetables	2 vegetables
2 protein sources	2 protein sources
2–3 dairy products—also supplies protein	2 dairy products—also supplies protein

The foods listed under "miscellaneous" and "beverages" may be used as desired to complete the daily diet. This grouping of foods allows the child and family to choose a variety of food items which they like and which will insure sound nutrition.

6. It is important to keep a diet diary for the days specified. This is the only method of control, as inadvertent infractions can be detected, and this allows us to monitor your intake for nutrient adequacy.

7. You are asked to keep a daily record of infractions. We would hope that none occur. However, should infractions occur, it is important that you list them all.

Diet N

These foods are to be ELIMINATED from diet:

Cereals and grain products
All breakfast cereals with artificial colors or flavors
All cakes, cookies, pastries, sweet rolls, doughnuts, breads, etc., with artificial flavors or colors (i.e., from bakery)
Manufactured pie crusts
Frozen baked mixes
Prepared poultry stuffing

Fruits
Almonds
Apples
Apricots
Berries—blackberries, blueberries, boysenberries, gooseberries, raspberries, strawberries
Cherries
Currants—grapes and raisins or any products made of grapes, (e.g., wine, wine vinegar, jellies, etc.)
Nectarines
Oranges
Peaches

Vegetables
Tomatoes and all tomato products
Cucumbers (pickles)

Protein sources
Meats
Bologna, luncheon meats*
Salami
Frankfurters
Sausage*
Meat loaf*
Ham, bacon, pork*
All barbecued types of chicken
All turkey prepared with basting, called "self-basting"
Frozen fish fillets that are dyed or flavored—fish sticks or patties, typed
   or flavored.

Dairy products
Manufactured ice cream or ice milk unless label specifies no synthetic
   coloring or flavoring
Colored cheeses (i.e., processed or yellow-orange)
All instant breakfast drinks and preparations
Flavored yogurt
Prepared chocolate milk
Colored butter

Beverages
Cider
Wine
Beer
Diet drinks
Tea, hot or cold
All carbonated beverages except 7 UP

Miscellaneous
Sherbets, ices, gelatins, junkets, puddings with artificial flavor or color-
   ing
Powdered pudding, Jell-o, and drink mixes
All dessert mixes
All manufactured candy—hard or soft
Oleomargarine
Prepared mustard

*When flavored or colored (usually indicated on package).

All mint flavored and wintergreen flavored items
Gum
Oil of wintergreen
Cloves
Jam or jellies made with artificial colors or flavors and fruits not allowed
Soy sauce, if flavored or colored
Cider vinegar
Wine vinegar
Commercial chocolate syrup
Barbecue-flavored potato chips
Catsup
Chili Sauce

Sundry items
Aspirin, Bufferin, Excedrin, Alka-Seltzer, Empirin, Empirin Compound, Anacin
Vitamins
All toothpastes and toothpowder*
All mouthwashes
All cough drops
All throat lozenges
Antacid tablets
Perfumes

These foods are to be PERMITTED on diet:

Cereal and grain products
Any cereal without artificial colors or flavors, dry or cooked
Any baked goods without artificial color or flavor—most baked goods will have to be prepared at home
All flours
All commercial breads except egg brand (usually dyed)

Fruits
Grapefruit†
Lemon†
Lime†

---

*A salt-and-soda mixture can be used for cleaning teeth. Neutrogena soap (unscented) can be substituted for toothpaste or powder.

†Good sources of vitamin C, juices may be labeled "vitamin C enriched."

Pear
Pineapple
Banana
Juices of these fruits without artificial flavor or color

Vegetables
All vegetables except tomatoes and cucumber (some vegetable mixtures have artificial color or flavor – these are not allowed)
Potatoes in any form

Main protein sources
Meats
All meats not listed on elimination list
All poultry
All fish not flavored or colored

Legumes
All dried peas, beans, peanuts except those colored or flavored
All fresh eggs

Dairy products
Milk
All natural cheese
Plain yogurt
Vanilla and chocolate ice cream made with natural ingredients
Butter, not colored

Beverages
Homemade lemonade
7 UP
Coffee

Miscellaneous
Dry mustard
Jam, jellies made from permitted fruits, not artificially colored or flavored
Honey
Real mayonnaise
Distilled white vinegar
Homemade candies, without almonds
Homemade chocolate syrup – for all purposes

Instructions for Management with the M Diet (Control Diet)

1. The diet lists all foods and products that are to be prohibited.

2. Any item not on the list is permitted. It is necessary to check every food package or container very carefully for chocolate or unenriched flour. In many cases, type of flour is not indicated. If the label does not read "enriched flour", do not use it. When in doubt, do not experiment—eliminate the food item.

3. With the difficulty in obtaining many food items on the market that conform to the allowed food list, it will be necessary to prepare many foods at home. For example, food items that are normally sold prefried should be cooked a different way, and chocolate-containing foods should be prepared only with instant chocolate.

4. The greatest success is usually observed when the entire family follows the diet. This removes the temptations from the child and facilitates compliance with the diet.

5. It is important that the child and family continue maintaining adequate nutrition during the study. For this reason, the foods on the allowed list have been grouped according to their contribution to the diet. A balanced diet which will maintain the nutritional status of the individuals following the diet can be achieved by choosing a certain number of foods from each group, to be eaten each day. The minimum number of servings from each group is suggested as follows:

Child	Adult
4 cereal and grains	4 cereal and grains
2 fruits (1 good source of vitamin C)	2 fruits (1 good source of vitamin C)
2 vegetables	2 vegetables
2 protein sources	2 protein sources
2–3 dairy products	2 dairy products

The foods listed under "miscellaneous" and "beverages" may be used as desired to complete the daily diet. This grouping of foods allows the child and family to choose a variety of food items which they like and which will insure sound nutrition.

6. It is important to keep a diet diary for the days specified. This is the only method of control, as inadvertent infractions can be detected, and this allows us to monitor the child's intake for nutrient adequacy.

7. You are asked to keep a daily record of infractions. We would hope that none occur. However, should infractions occur, it is important that you list them all.

Diet M

These foods are PERMITTED on the diet:

Cereal and grain products
All cold cereal with milk and without chocolate
Converted or long-cooking rice
All bread products made with enriched white flour or whole wheat, etc.
Pastries without cream filling, made with enriched flour
Cakes and cookies without chocolate

Fruits
All fruits and their juices except those on elimination list
Oranges*
Strawberries*

Vegetables
All vegetables not on elimination list
(corn, beans, tomatoes†, broccoli†, carrots††)

Main protein sources

Meat
Meats not on elimination list
Luncheon meats, not containing by-products or ingredients on elimination list
Chicken and turkey without skin

Legumes
Dried peas and beans, lentils
Baked beans without pork
Eggs—fresh or frozen but not fried

*Good sources of vitamin C.
†Good sources of vitamins A and C.
††Good sources of vitamin A.

Dairy products
Milk
Yellow cheeses and cheese food
Pudding, custards, ice cream not containing chocolate or unenriched
    flour

Miscellaneous
Pretzels
Cheese–corn snacks
Dessert toppings and syrups
Soup without noodles or eliminated vegetables
Baker's yeast
Wine and cider vinegar
Candy without chocolate
Margarine

Beverages
Cola drinks
Fruit-flavored drinks with allowed fruits
Tea

Diet M

These foods are to be ELIMINATED from diet:

All fried foods—foods cooked in fat or oil (breaded or greasy foods)

Cereal and grain products

All hot (cooked) cereals (Maypo, oatmeal, farina)
Waffles
Pancakes
French toast
All crackers (graham, saltines, cheese)
Corn bread
Instant rice
Cream-filled pastries
Chocolate cakes and cookies
All foods containing unenriched white flour

Fruits
Grapefruit
Lemons
Melons
Bananas
Walnuts

Vegetables
Mushrooms
Lettuce
Peas
Panfried or french fried potatoes

Main protein sources

Meat
Corned beef
Ham, pork, bacon
Fresh fish (frozen allowed)
Tuna fish
Chicken skin and turkey skin
Giblets

Legumes
Peanuts
Peanut butter

Dairy products
Butter (margarine allowed)
White cheese
Chocolate pudding
Chocolate ice cream

Beverages
Root beer
Ginger ale
Lemonade
Coffee
Chocolate milk

Miscellaneous
Pickles in cloudy liquid
Popcorn

Chocolate syrup (instant chocolate, dry powder is acceptable)
Soup with noodles
White cane sugar
Brewer's yeast
Chocolate candy
Distilled vinegar
Honey

Sundry items
Chewing gum except for sugarless gum
Deodorant soap
Bubble bath
Lemon shampoo
Eye drops (except doctor's prescription)
Egg-based paints
Gummed envelopes, stamps, stickers, tapes

Frequency

Name _____

Subject number _____

Condition _____

Day _____

Week of _____

	N	–	1	2	3	4	5	6	7
X/ month									
1. Doughnuts									
2. Sugar-coated cereals									
3. Oatmeal									
4. Home-made bread and rolls									
5. Store-bought bread and rolls									
6. "Natural" bread and rolls									
7. Graham crackers									
8. Store-bought cookies									
9. Chocolate cake									
10. Cake from a mix									
11. Macaroni, spaghetti, noodles									
12. Oranges and orange juice									
13. Grapefruit and grapefuit juice									
14. Apples, juice and cider									
15. Bananas									
16. Tomato and tomato products									
17. Canned fruits									
18. Peas									
19. Green beans									

Figure 30. Weekly checklist to be filled out by parent indicating frequency of consumption of specific foods.

Frequency (Continued)

X/
month

	N	–	1	2	3	4	5	6	7
20. Corn and corn products									
21. Carrots									
22. Frozen vegetable combinations									
23. Potatoes									
24. Peanut butter									
25. Hamburger									
26. Pork or ham									
27. Luncheon meats									
28. Roasts									
29. Chicken									
30. Tuna fish									
31. Frozen fish sticks									
32. Butter									
33. Whole milk									
34. Chocolate milk									
35. Ice cream									
36. American cheese									
37. Cottage cheese									
38. TV dinners									
39. Margarine									
40. Sherbet									
41. Canned pudding									
42. Dinner helpers									
43. Home-made casseroles									

(Continued)

Frequency (*Continued*)

X/
month

	N	–	1	2	3	4	5	6	7
44. Canned soup									
45. Sandwiches									
46. Pancakes									
47. Plain dry cereal									
48. Soft drinks									
49. Chocolate candy									
50. Non-chocolate snack food									

Time	Place	Food	Description or method of preparation	Amount	Code

Diet Record

Name _____ Condition _____ Day _____

Subject number _____ _____ Week: 1 2 3 4

Figure 31. Complete diet record of all food consumed including where and when eaten and method of preparation.

List of Infractions

Child's name _____

Diet assigned _____

Week of the _____

Date	Food item — description brand name	Amount	Why is this item considered an infraction?	Behavior observed

Figure 32. A complete recording of any infractions of the diet must be kept.

Dietary Questionnaire

Name _____ Age _____ Date _____
1. Usual meal times _____
   (Sunday _____)
2. Between meal time: A.M. _____ P.M. _____ before bed _____
   Usually what? _____
3. Appetite: Good _____ Varies _____ Poor _____
4. Food dislikes _____
   Will he/she try them? _____
5. Food likes (10) _____
   _____
6. Supplements—what and how often _____
7. Problems at meal (table with child) _____
   _____
   _____
8. Meals at school (how and what) _____
9. Any medical complications with food? _____
10. Are there any changes in his/her food habits when the child is not on medication? ____
   _____
   _____

Comments: _____
_____
_____
_____

Dietary Degree of Difficulty Questionnaire

Scoring:
Patient's name _____ a = 0
Patient's number _____ b = 1
Diet code _____ c = 2
d = 3
e = 4

1. How much extra shopping time was required to follow this diet?
   a. None
   b. Just a little
   c. Pretty much
   d. Very much
2. How much extra cost was required to follow this diet?
   a. None
   b. Just a little
   c. Pretty much
   d. Very much

3. How much extra time was required at home to prepare meals as a result of this diet?
   a. None
   b. Just a little
   c. Pretty much
   d. Very much

4. Compared to your child's "normal" diet, how much food was consumed during an average meal with this special diet?
   a. Much more
   b. A little more
   c. No difference
   d. A little less
   e. Much less

5. How many of your son's/daughter's favorite foods were eliminated as a result of this diet?
   a. None
   b. Just a few
   c. A moderate number
   d. A large number

6. How often did your son/daughter complain about food items *not* available as a result of this diet?
   a. Not at all
   b. Just a little
   c. Pretty much
   d. Very much

7. On the average, how often did your son/daughter consume "forbidden" food items?
   a. 0–1 times/week
   b. 2–3 times/week
   c. 4–5 times/week
   d. Over 5 times/week

8. In general, how difficult was it to comply with this diet?
   a. Not at all
   b. Just a little
   c. Pretty much
   d. Very much

9. Please note any other problems or difficulties you have had in following this diet.
   _____
   _____
   _____
   _____
   _____

HYPERKINESIS STUDY DIET

Instructions

1. The diet eliminates all foods and beverages containing artificial colors and artificial flavors.

2. The diet does not eliminate food preservatives, natural salicylates, or any other food additives except artificial colors and artificial flavors. However, if the child has a known or suspected food allergy, that food should be avoided. This information should be indicated on the Three-Day Food Record.

3. A market survey was conducted in the Greater Pittsburgh area in which package labels were checked and, in some cases, food companies were contacted to determine if artificial colors and/or artificial flavors were present. The following pages contain foods found to be free of artificial colors and flavors; these are listed under the "Allowed" column. These foods may be eaten as desired. The foods that were checked and found to contain artificial colors and/or artificial flavors are listed under the "Avoid" column. These foods, or any food containing the "Avoid" foods, should not be eaten.

4. Because ingredients in products may change, and because new products are continually entering the market, it is necessary to check every food package or container very carefully for artificial colors and flavors. Foods should be avoided if the following words appear on the label: "artificial color," "U.S. certified color," "FD&C colors," "artificial flavor," and/or "imitation flavor." When in doubt, eliminate the food item. If you become aware of any change in ingredients which alters the listing of the food, please let us know.

5. Carotene, beta-carotene, malt flavoring, and caramel coloring are natural food additives and are permitted.

Note

This diet was developed for the Hyperkinesis Study and, as such, should be considered preliminary and subject to change. Therefore, it is recommended that the diet be followed only under the supervision of our pediatrician and nutritionist.

Hyperkinesis Study Diet

Allowed	Avoid
White milk	Flavored milk
Eggs	Egg substitutes (Eggbeaters, etc.)
Cheese	Cheese
All Kraft natural cheeses (Cheddar, Brick, Swiss, Colby, Muenster, Monterey Jack)	American cheese (all brands) All other processed cheese food
Mozareella (all brands)	
Ricotta (all brands)	
Parmesan (all brands)	
Cottage cheese (Sealtest, Otto, Breakstone, Trim, Menzie)	All other brands of cottage cheese
Farmer cheese (Breakstone)	
Cream cheese	
Nabisco Snack-mate cheese spread (Zesty Caraway only)	All other cheese spreads
Butter	
Ice cream (Breyer's)	Margarine (all brands) All other brands of ice cream Sherbert
Sour cream (Breakstone's)	All other brands of sour cream
Yogurt (Dannon's)	All other brands of yogurt
Meat, fish, poultry	
Fresh meat, fish, poultry	
Processed packaged meats (e.g., lunch meats, sausage, skinless weiners, bacon)	
Oscar Mayer	Weiners with casing
Sugardale	Delicatessen hot dogs
Armour	
Sinai	
Frozen fish (Mrs. Paul's)	Frozen self-basting chicken
Frozen meat products (Jiffy)	Frozen turkey and chicken roll (loaves)
Frozen dinners (Swanson)	

Allowed	Avoid
Canned tuna, salmon, Spam	
Canned sandwich spreads (Underwood)	
Canned pork luncheon meat (Cliff House)	
Breads, cereals, pastas	
Breads	Breads
Store-brand breads, buns, English muffins	Egg breads and egg rolls
Pepperidge Farm bread products	
Catherine Clark breads	
Pop Tarts (Sugar and Cinnamon only)	Pop Tarts (all other varieties)
Pancakes and waffles (Aunt Jemima)	Frozen waffles
Buckwheat and Whole Wheat Pancake Mixes, Pillsbury Hungry Jack	All other types and brands of pancake and waffle mixes
Extra Light Pancake and Waffle Mix, Kroger Buttermilk Pancake Mix)	
Hot roll mix (Pillsbury, Duff's)	
Corn muffin mix (Flako, Dromedary)	
Bisquick	
Frozen bread dough	All other roll mixes
Stuffing mix (Stove Top chicken flavor and stuffing with rice *only*)	All other stuffing mixes
Refrigerator rolls (Pillsbury Crescent, Butterflake, and 1869 Biscuits *only*)	All other refrigerator biscuits and rolls
Cornflake crumbs (Kellogg's)	Packaged breading mixes (Shake 'n' Bake)
Pastas	Pastas. If artificial colors or flavors are contained in the product, they *must* be listed on the package, and these products must be avoided.
Noodles, spaghetti, macaroni	Rice-a-Roni Fried Rice with Almonds
Rice (white, brown, wild rice)	
Rice-a-Roni (all flavors except under "Avoid")	
Uncle Ben's Rice mixes (all flavors except under "Avoid")	Uncle Ben's Fast Cooking Long Grain and Wild Rice
Cereals	Cereals
Kellogg's Raisin Bran	All other cereals
Kellogg's Rice Krispies	

(Continued)

Allowed	Avoid
Kellogg's Corn Flakes	
Kellogg's Country Morning	
Kellogg's Shredded Wheat	
Kellogg's Frosted Mini Wheats	
Kellogg's Bran Buds	
Kellogg's Sugar Frosted Flakes	
General Mills Kix	
Quaker Puffed Rice	
Quaker Puffed Wheat, 100% Natural	
Quaker Peanut Butter Cap'n Crunch	
Nabisco Team	
Ralston Purina Corn Chex	
Ralston Purina Wheat Chex	
Ralston Purina Rice Chex	
Heartland Natural cereal	
Post Raisin Bran	
Post Grape Nuts Flakes	
Post Grape Nuts	
Post Frosted Rice Krinkles	
Post 40% or 100% Bran	
Post Sugar Crips	
Nature Valley Granola	
Hot cereals: Cream of Wheat, Farina	
Wheatena, Instant Ralston, Oatmeal, Cream of Rice, Quaker Ready-to Serve Oatmeal (regular and cinnamon spice *only*)	
Fruits and vegetables	
Fresh, canned, dried, frozen fruits	Orange peels
	Fruit cocktail
	Maraschino cherries
Fresh, canned, frozen vegetables	Frozen vegetables with sauces
Frozen vegetables without sauces or seasonings	Frozen vegetables in boiling bag
Mrs. Paul's frozen vegetables (onion rings, zucchini, eggplant, french fries)	
Instant potatoes (Betty Crocker's, French's)	French's Potatoes au Gratin
Desserts	
Cookies	
Nabisco Butter Flavored Cookies	All other cookies
Nabisco Peanut Creme Patties	
Nabisco Raisin Fruit Biscuits	
Sunshine Vanilla Wafers	

Allowed	Avoid
Sunshine Ginger Snaps	
Keebler Old Fashioned Oatmeal	
Keebler Pitter Patter Peanut	
Keebler Butter Creme	
Pepperidge Farm: All cookies	
FFV Lemon Thins	
Homemade with allowed ingredients	
Pies	
Pie crust mix (Jiffy)	
Pie fillings	Commercial and bakery pies
Thank You Brand apple, apricot, blackberry, raisin	All other flavors and brands of pie filling and crust mixes
Lucky Leaf apple, raisin	
Frozen pies	
Mrs. Smith Pumpkin Custard	
Mrs. Smith Cherry Pie	
Morton Pumpkin Pie	
Homemade with allowed ingredients	
Cakes and mixes	
Pepperidge Farm cakes	All canned and packaged frosting mixes
Betty Crocker Snackin' Cake (applesauce raisin)	
Betty Crocker Gingerbread Mix	
Pillsbury Apple Cinnamon Coffeecake	
Pillsbury Cinnamon Streusel Coffeecake	
Pillsbury Gingerbread Mix	
Pillsbury Date Quick Bread	
Pillsbury Nut Quick Bread	
Pillsbury Apricot Nut Bread	
Aunt Jemima Easy Coffeecake	
Washington Old Fashioned Spice Cake	
Sara Lee Original Cheese Cake	
Sara Lee Chocolate Pound Cake	
Sara Lee French Crumb Cake	
Sara Lee Pecan Coffee Cake	
Sara Lee Caramel Pecan Rolls	
Homemade with allowed ingredients	
Other desserts	
Knox unflavored gelatin	Flavored packaged gelatin (Jell-o, Royal, etc.)

(Continued)

Allowed	Avoid
Minute Tapioca	Pudding mixes
	Canned pudding
Pepperidge Farm frozen desserts (except those containing margarine)	All other frozen desserts
	Ice cream toppings
	Whipping toppings (canned, frozen, packaged)
	Ice cream cones
Candy	
Planter's Peanut Bars	All other candy
Reese's Peanut Butter Cups	Marshmallows and Mallo Fluff
Cracker Jacks	
Fiddle Faddle	
Kraft Peanut Brittle	
Crunchola Bars (cinnamon *only*)	
Life Savers (CrystoMint, Spearmint, PeppoMint *only*)	All other Life Savers
Wrigley's Spearmint and Doublemint	All other chewing gum
Beverages	
Fresh or frozen fruit juices	Orange Plus, Awake
Frozen lemonade (Minute Maid)	Pink lemonade
Lemon Blend	Fruit flavored drinks
Lemon juice	Diet drinks
Nectars	Kool Ade
Coffee	Packaged drink mixes
Tea	Coffee creamers
Instant (Nestea with sugar and lemon, plain—no sugar lemon)	All other brands of tea mix
Coca-Cola, Pepsi, RC Cola	
Gingerale, 7-Up	
Sprite	All other soda pop
Commercial packaged products	
Canned soup	Dry packed soup mixes
Campbell (all soups not specifically listed under AVOID)	Campbell soups: Green pea, cream of chicken, cream of celery, cream of mushroom, cream of potato, cream of onion, chicken noodle, chicken alphabet, chicken with rice, chicken and dumplings, cheddar cheese, golden mushroom
Red & White (all soups not containing margarine or other AVOID ingredients)	
Cream of mushroom, cream of celery, cream of potato, chicken noodle	

Allowed	Avoid
All other soups not containing margarine or other AVOID ingredients	Any soup containing margarine or other AVOID ingredients
Frozen pizza (Jeno's, Tony's, Totino)	
Ragu Spaghetti Sauce	
Chef Boy-ar-dee products	
Hamburger Helper (Potato Stroganoff and Chili Tomato *only*)	
Old ElPaso Mexican canned foods	
La Choy Chinese canned foods	
Franco-American Spaghettios, Beef Ravioli, Spaghetti, Macaroni and Cheese, Macaroni and Meatballs	
Heinz Spaghetti in Tomato Sauce with Cheese	

Snacks and crackers Snacks

Allowed	Avoid
Pretzels (check labels)	All cheese-flavored snacks
Nuts	
Popcorn	Commercial popcorn
Sunflower seeds	
Soybeans	
Potato chips (plain)	Barbequed potato chips
Fritos (plain)	Flavored Fritos

Crackers

Allowed	Avoid
Saltines (all brands)	
Nabisco Ritz	All other crackers
Nabisco Triscuits	
Nabisco Sociables	
Nabisco Waverly Wafers	
Nabisco Chicken-in-a-Biscuit	
Nabisco Swiss Cheese	
Nabisco Buttery Sesame	
Nabisco Bacon	
Nabisco Zwieback	
Nabisco Arrowroot Biscuits	
Royal Lunch Milk Crackers	
FFV Stoned Wheat Wafers	Wyler's Beef Bouillon
FFV Roman Meal Wafers	
Sunshine Hi-Ho's	
Sunshine Country Cheddar N' Sesame seed Oysterettes	

Miscellaneous

Baking powder and baking soda
Bouillon and broth
    Wyler's Chicken Bouillon
    College Inn Beef Broth

(Continued)

Allowed	Avoid
College Inn Chicken Broth	
Chocolate	
Baker's German Sweet Chocolate	
Baker's Unsweetened Chocolate	
Hershey's Cocoa	
Hershey's Baking Chocolate	
Coconut	
Flour	
Gravy Master	
Jelly, honey, molasses	
Ketchup (Heinz, Hunt's), Mustard	
Mayonnaise (Hellman's Kraft)	
Olives (Food Club, Hefty Fair)	
Peanut Butter	
Pickles	
Heinz Genuine Dills	Pickled red peppers
Heinz Dills	All other types and brands of pickles
Heinz Polish Dill Spears	
Heinz Sweet Pickles	
Heinz Sweet Cucumber Slices	
Heinz Sweet Gherkins	
Heinz Sweet Onions	
Vlasic Genuine Dills	
Vlasic Cocktail Onions	
Food Club Dills	
Food Club Sweet Relish	
Salad Dressings	
Kraft Miracle Whip	Kraft Golden Caesar
Kraft (all except those under AVOID)	Kraft Green Onion
	Kraft Creamy Cucumber
Seven Seas Herbs and Spices	Kraft low calorie dressings
Dutch Pantry Sweet and Sour	All other salad dressings
Dutch Creamy Sweet and Sour	
Marzetti Thousand Island	
Marzetti Country French	
Marzetti Slaw Dressing	
Marzetti Blendaise	
Sauces	
Heinz 57	Barbeque Sauces
A–1 Steak Sauce	
Shortening and cooking oils	
Spices and herbs (all)	
Syrups	
Heartland *Natural* Syrup	All other pancake and waffle syrups
Log Cabin *Natural* Syrup	
Daily's *Natural* Syrup	
Sugar (granulated, brown, powdered)	
Vanilla (natural extract)	Vanillin

Allowed	Avoid
Vinegar (white, cider, wine)	
Yeast	

Health aids

Toothpaste (Colgate, Pepsodent, Ul-trabrite)	All other brands of toothpaste
	All other brands of mouthwash
Mouthwash (Listerine, Cepacol)	Cough drops and lozenges
	Vitamins (colored and flavored)
	Medicines (colored and flavored)

Fast foods

Information received from MacDonald's fast food restaurants indicates that the following foods contain artificial colors and artificial flavors and, therefore, must be AVOIDED:

Fish	Pancakes
Pies	Pancake syrup
Cookies	Colored soda pop
Pickles	Coffee creamer
Milkshakes	Margarine

A hamburger is ALLOWED if the pickle is removed. All other foods served at MacDonald's (e.g., french fries, sauce on hamburgers) are ALLOWED.

THE CYTOTOXIC FOOD TEST

Purpose

Cytotoxic testing is a controversial *in vitro* diagnostic technique used to determine all allergies to foods and chemicals. The test is based on the observation that living leucocytes undergo destruction as a result of the food antigen–antibody reaction. Since the application is no more than a simple bioassay, such a procedure is extremely practical because, for the subject, it requires only the time to draw a blood sample, and yet it permits the testing of a large number of suspected agents. However, the validity of cytotoxic testing as a diagnostic tool has not been assessed adequately, thus resulting in much debate over the utility of the procedure.

The purpose of the present investigation is to determine if leucocytosis occurs to artificial colors when cytotoxic testing is performed on hyperactive children who have demonstrated behavioral improvement when placed on the KP dietary regime. Since these children are suspected of having a sensitivity to one or more artificial colors, cytotoxic testing should reveal positive results.

Preparation of Glassware

The cytotoxic food test starts with the preparation of glassware. The glassware involved with actual testing are:

Microscope Slides   Test Tubes
Antigen Dropper Bottles   Coverslips

Pasteur Pipettes   Porcelain Dishes
Bryan Tubes

All glassware is prepared with the following solutions:

NaOH solution (2 h)
NCL Acid Alcohol (2 h)

Each slide, bottle, test tube, pipette, and coverslip is individually rinsed.
Coverslips and microscope slides are individually dipped in alcohol and
blotted dry. (Bryan technique: Rinse all coverslips and dry together and
not separately. Also slides and coverslips are all reused in the Bryan
Technique.)

The microscope slides are fixed with three vaseline rings which are
used for reservoirs for food allergen. Next, equal parts of powdered food
antigen and sterile distilled water are mixed in antigen bottles and left to
settle for 24 h. The supernatant is then used in a 1:5 ratio with sterile
distilled water. Then the reservoirs on microscope slides are filled with
solution. Slides sit covered 12–24 h until dry.

Preparation of Blood Sample

The second part of the Cytotoxic Food Test is the collection of a
blood sample and the preparation of blood.

4 cc of venous blood is drawn from patient and mixed immediately
with citrate and placed in Bryan tubes. (Bryan Technique: 10cc of blood
are drawn and placed in at least five tubes instead of just two tubes.)

The white blood cell count is prepared.

Blood is spun at a speed of 25 rpm for ½ h to 45 min. (Bryan Technique: blood is spun at low speed for 1 h.) Serum is suctioned from
Bryan Tubes with pasteur pipette and mixed with sterile distilled water
in a 1:4 ratio. White cells are then suctioned from Bryan Tubes and mixed
with serum solution. White blood cell (WBC) mixture is fixed on microscope slide and covered with coverslips.

Preparation is set aside for two or more h.

WBC count solution is fixed on hemocytometer and count is taken.

Reading Slides

The third part of the cytotoxic food test is reading the slides. Each reservoir is now read under microscope and for each food, cells in ten fields are read. Results are then recorded as follows:

Slight reaction—¼ of cells appear immobile or cells are vacuolated.

Moderate reaction—½ of cells appear immobile and cells are vacuolated.

Marked reaction—¾ of all WBC to all WBC are immobile and vacuolated. Ghost cells may be present.

OBSERVATIONAL RATING SCALES[1]

I.   Gross Motor Activity

    1. Hypoactive (sluggish, lacks spontaneous motor activity)

    2. Average in gross motor activity (No excessive running, no restlessness, or fidgety behavior)

    3. Restless and fidgety; can sit for appropriate periods of time, but squirms in chair, moves about in chair, does not sit still

    4. Hyperactive, but can be controlled; has difficulty sitting down; gets up, but can be brought back

    5. Severe hyperactivity; cannot be controlled; runs around, climbs, cannot sit for any length of time

II. Distractibility and Concentration (Only to be rated for tasks which are presented to the child, not those which he/she does spontaneously)

    1. Is goal-oriented, maintains interests in tasks

    2. Is slightly distractible, but goal-oriented and will finish tasks with minimal pressure

    3. Distractible; finishes tasks only with considerable pressure; demonstrates some goal-oriented behavior

    4. Quite distractible; has difficulty completing task regardless of external pressure; shows minimal goal-oriented behavior

    5. Extremely distractible, cannot sustain attention regardless of external pressure, not goal-oriented

---

[1]These scales are a slight modification of scales recommended by Dr. Rachel Gittelman in personal communication to the author's staff. We are indebted to Dr. Gittelman for allowing us to use these scales.

III. Frustration Tolerance

    1. Underreacts to frustrating limit setting; lack of any reaction indifferent to limits

    2. Average reaction to frustration; can accept setting of limits without showing either indifference or emotional upset

    3. Has difficulty relinquishing demands; persists when wants something; may whine and carry on verbally, but no emotional upset

    4. Exaggerated reaction to frustration, but stops after a short time; may cry, scream, have a brief tantrum

    5. Catastrophic reaction to even minor frustrations; has violent tantrums.

IV. Mood

    1. Child is happy and cheerful

    2. Child is mostly happy and cheerful, but may briefly appear gloomy

    3. _____

    4. Child is mostly gloomy and depressed

    5. Child is gloomy and depressed

V. Attention Seeking Devices. Please consider so-called "negative" attention-seeking behaviors like the following: needless requests or questions, silly verbal behavior, clowning, showing off, shouting testing limits, tattling, crying, tantrums, hiding, playing sick or other. (Apart from your rating, please mention what kind of attention-seeking devices this child employs.)

    1. Child never seeks attention through devices similar to the ones described above.

    2. _____

    3. Child occasionally employs such devices

    4. _____

    5. Child quite frequently resorts to such devices

VI. Impulse Control

    1. Child has good control of his/her impulses; very seldom acts impulsively.

    2. _____

    3. Child has some control of his/her impulses, but sometimes act rather impulsively.

    4. _____

    5. Child is extremely impulsive, seldom stops to think about the consequences of his/her actions.

REFERENCES

Baldwin, D. G., Kittler, F. J., & Ramsay, R. G. The relationship of allergy to cerebral dysfunction. *Southern Medical Journal*, 1968, **61**, 1039–1041.

Beal, V. The nutritional history in longitudinal research. *Journal of the American Dietetic Association*, 1967, **51**, 426–432.

Benson, T. E., & Arkins, J. A. Cytotoxic testing for food allergy evaluation of reproducibility and correlation. *Journal of Allergy and Clinical Immunology*, 1976, **58**, 471–476.

Bernstein, J. E., Page, J. G., & Janicki, R. S. Some characteristics of children with minimal brain dysfunction. In C. K. Conners (Ed.), *Clinical use of stimulant drugs in children*. Netherlands: Excerpta Medica, 1974, pp. 24–35.

Black, A. P. A new diagnostic method in allergic disease. *Pediatrics*, 1956, **17**, 716–724.

Bosco, J. J., & Robin, S. S. Hyperkinesis: How common is it and how is it treated? In C. H. Whalen & B. Henker (Eds.), *Hyperactive children: The social ecology of identification and treatment*. New York: Academic, 1979.

Bryan, W. T. K., & Bryan, M. P. The application of in vitro cytotoxic reactions to clinical diagnosis of food allergy. *Larnygoscope*, 1960, **70**, 810–824.

Bryan, W. T. K., & Bryan, M. P. Cytotoxic reactions in the diagnosis of food allergy. *Laryngoscope*, 1969, **79**, 1453–1472.

Bryan, W. T. K., & Bryan, M. P. Cytotoxic reactions in the diagnosis of food allergy. *Otolaryngological Clinics of North American*, 1971, **4**, 523–534.

Bryan, W. T. K., & Bryan, M. P. Clinical examples of resolution of some idiopathic and other chronic disease by careful allergic management. *Laryngoscope*, 1972, **82**, 1231–1238.

Campbell, M. B. Neurologic manifestations of allergic disease. *Annals of Allergy*, 1973, **31**, 485–498.

Cantwell, D. P. Psychiatric illness in the families of hyperactive children. *Archives of General Psychiatry*, 1972, **27**, 414–417.

Cantwell, D. P. Epidemiology, clinical picture and classification of the hyperactive child syndrome. In D. P. Cantwell (Ed.), *The hyperactive child*. New York: Spectrum, 1975, 3–15.

Chalmers, F. W., Clayton, M. M., Gates, L. O., Tucker, R. E., Wertz, A. W., Young, C. M., & Foster, W. D. The dietary record, how many and which days. *Journal of the American Dietetic Association*, 1952, **28**, 711–717.

Clarke, T. W. The relation of allergy to childhood neuroses. *Journal of Child Psychiatry*, 1948, **1**, 177–180.

Congressional Record. Senate, October 3, 1973, §19736–§19742.

Conners, C. K. Rating scales for use in drug studies with children. *Psychopharmacology Bulletin, Special Issue on Children*, 1973, 24–42. (a)

Conners, C. K. Psychological assessment of children with minimal brain dysfunction. *Annals of the New York Academy of Sciences*, 1973, **205**, 283–302. (b)

Conners, C. K. Food additives and hyperactivity. In R. N. Knights & D. Bakker (Eds.), *Rehabilitation, diagnosis and treatment of learning disabilities*. New York: University Park Press, in press.

Conners, C. K., & Eisenberg, L. The effects of methylphenidate on symptomatology and learning in disturbed children. *American Journal of Psychiatry*, 1963, **120**, 458–464.

Conners, C. K., & Goyette, C. H. The effect of certified food dyes on behavior: A challenge test. *New Clinical Drug Evaluation Unit Intercom*, 1977, **7**, 18–19.

Conners, C. K., & Werry, J. S. Pharmacotherapy. In H. C. Quay & J. S. Werry (Eds.), *Psychopathological disorders of childhood* (2nd ed.). New York: Wiley, 1979, pp. 336–386.

Conners, C. K., Eisenberg, L., & Barcai, A. Effect of dextroamphetamine on children: Studies on subjects with learning disabilities and school behavior problems. *Archives of General Psychiatry*, 1967, **17**, 478–485.

Conners, C. K., Eisenberg, L., & Sharpe, L. Effects of methylphenidate (Ritalin) on paired-associate learning and Porteus Maze performance in emotionally disturbed children. *Journal of Consulting Psychology*, 1964, **28**, 14–22.

Conners, C. K., Goyette, C. H., Southwick, D. A., Lees, J. M., & Andrulonis, P. A. Food additives and hyperkinesis: A controlled double-blind experiment. *Pediatrics*, 1976, **58**, 154–166.

Conners, C. K., Goyette, C. H., & Newman, E. B. *Dose–time effect of artificial colors in hyperactive children*. Paper presented at the annual meeting of the American Psychological Association, Toronto, Canada, August 18, 1978.

Crook, W. G., Harrison, W. W., Crawford, S. E., & Emerson, B. S. Systemic manifestations due to allergy. *Pediatrics*, 1961, **27**, 790–799.

David, O., Clark, J., & Voeller, K. Lead and hyperactivity. *Lancet*, 1972, **2**, 900–903.

David, O. J., Hoffman, S. P., Sverd, J., & Clark, J. Lead and hyperactivity: Lead levels among hyperactive children. *Journal of Abnormal Child Psychology*, 1977, **5**, 405–416.

Denson, R., Nanson, J. L., & McWatters, M. A. Hyperkinesis and maternal smoking. *Canadian Psychiatric Association Journal*, 1975, **20**, 183–187.

Eisenberg, L., & Conners, C. K. Psychopharmacology in childhood. In N. B. Talbot, J. Kagan, & L. Eisenberg (Eds.), *Behavioral science in pediatric medicine*. Philadelphia: Saunders, 1971.

Eisenberg, L., Gilbert, A., Cytryn, L., & Molling, P. A. The effectiveness of psychotherapy alone and in conjunction with perphenazine or placebo in the treatment of neurotic and hyperkinetic children. *American Journal of Psychiatry*, 1961, **117**, 1088–1093.

Feingold, B. F. Recognition of food additives as a cause of symptoms of allergy. *Annals of Allergy*, 1968, **26**, 309–313.

Feingold, B. F. *Introduction to clinical allergy*. Springfield, Ill.: Charles C Thomas, 1973.

Feingold, B. F. *Hyperkinesis and learning difficulties (H–LD) linked to the ingestion of artificial colors and flavors*. Paper presented at the annual meetings of the American Medical Association, section on allergy, Chicago, June 24, 1974.

Feingold, B. F. Hyperkinesis and learning disabilities linked to artificial food flavors and colors. *American Journal of Nursing*, 1975, **75**, 797–803. (a)

Feingold, B. F. *Why your child is hyperactive*. New York: Random House, 1975. (b)

Frazier, C. A. *Coping with food allergy*. New York: Quadrangle, 1974.

Galant, A., Zippin, C., Bullock, J., & Crisp, J. Allergy skin test: 1. Antihistamine inhibition. *Annals of Allergy*, 1972, **30**, 53–63.

Goyette, C. H., Conners, C. K., Petti, T. A., & Curtis, L. E. Effects of artificial colors on hyperkinetic children: A double-blind challenge study. *Psychopharmacology Bulletin*, 1978, **13**, 39–40.

Harley, J. P., & Matthews, C. G. Hyperactivity in children and food additives: Experimental investigations. In R. Knights & D. Bakker (Eds.), *Rehabilitation, diagnosis and treatment of learning disabilities*. New York: University Park Press, in press.

Harley, J. P., Ray, R. S., Tomasi, L., Eichman, P. L., Matthews, C. G., Chun, R., Cleeland, C. S., & Traisman, E. Hyperkinesis and food additives: Testing the Feingold hypothesis. *Pediatrics*, 1978, **61**, 818–828.

Hoobler, B. E. Some early symptoms suggesting protein sensitization in infancy. *American Journal of Diseases of Children*, 1916, **12**, 29.

Kittler, F. J., & Baldwin, D. G. The role of allergic factors in the child with minimal brain dysfunction. *Annals of Allergy*, 1970, **28**, 203–206.

Kupietz, S., Bialer, I., & Winsberg, B. A. A behavior rating scale for assessing improvement in behaviorally deviant children: A preliminary investigation. *American Journal of Psychiatry*, 1972, **128**, 1432–1436.

Levitt, E. E. The results of psychotherapy with children: An evaluation. *Journal of Consulting Psychology*, 1957, **21**, 186–189.

Levy, F., Dumbrell, S., Hobbes, G., Ryan, M., Wilton, N., & Woodhill, J. M. Hyperkinesis and diet: A double-blind cross-over trial with a tartrazine challenge. *Medical Journal of Australia*, 1977, **1**, 61–64.

Levy, F., & Hobbes, G. *Hyperkinesis and diet: A replication study*. Unpublished manuscript, Prince of Wales Children's Hospital, Randwich, N.S.W. 2031, Australia.

Lieberman, P., Crawford, L., Bjelland, J., Cannell, B., & Rice, M. Controlled study of the cytotoxic food test. *Journal of the American Medical Association*, 1975, **231**, 728–730.

Lietze, A. Immunological aspects of food allergy. In A. H. Rowe & A. Rowe, Jr. (Eds.), *Food allergy: Its manifestations and control and the elimination diets—A compendium*. Springfield, Ill.: Charles C Thomas, 1972, pp. 520–527.

May, C. D. Food allergy: A commentary. *Pediatric Clinics of North America*, 1975, **22**, 217–220.

McPartland, R. J., Foster, F. G., Matthews, G., Coble, P., & Kupfer, D. The LSI movement activated recording monitor—An instrument to study motor rhythms. *Sleep Research*, 1975, **4**, 261. (abstract)

Morrison, J. R., & Stewart, M. A. A family study of the hyperactive child syndrome. *Biological Psychiatry*, 1971, **3**, 189–195.

Moyer, K. E. Allergy and aggression: The physiology of violence. *Psychology Today*, 1975, **9**, 76–79.

National Advisory Committee on Hyperkinesis and Food Additives. *Report to the Nutrition Foundation*. The Nutrition Foundation, Inc., 482 Fifth Ave., New York, N.Y., 1975.

National Institute of Mental Health. *Psychopharmacology Bulletin, Special Issue, Pharmacotherapy of Children*. 1973, 139–140.

O'Leary, K. D., Rosenbaum, A., & Hughes, P. C. Fluorescent lighting: A purported source of hyperactive behavior. *Journal of Abnormal Child Psychology*, 1978, **6**, 285–289.

Palmer, S., Rapoport, J. L., & Quinn, P. O. Food additives and hyperactivity. *Clinical Pediatrics*, 1975, **14**, 956–959.

Pasamanick, B., & Knobloch, H. Brain damage and reproductive casualty. *American Journal of Orthopsychiatry*, 1960, **30**, 298–305.

Randolph, T. G. Allergy as a causative factor of fatigue, irritability and behavior problems of children. *Journal of Pediatrics*, 1947, **31**, 560–572.

Rinkel, H. J., Lee, C. H., Brown, D. W., Willoughby, J. W., & Williams, J. M. The diagnosis of food allergy. *Archives of Otolaryngology*, 1964, **79**, 71–79.

Robins, L. N. Follow-up studies. In H. Quay & J. Werry (Eds.), *Psychopathological disorders of childhood*. New York: Wiley, 1979, pp. 483–514.

Rowe, A. H., & Rowe, A. H., Jr. Diagnosis of allergy. In A. H. Rowe & A. H. Rowe, Jr. (Ed.), *Food allergy: Its manifestations and control and the elimination diets—A compendium*. Springfield, Ill.: Charles C Thomas, 1972, pp. 23–40.

Rutter, M., & Graham, P. The reliability and validity of the psychiatric assessment of the child: I. Interview with the child. *British Journal of Psychiatry*, 1968, **114**, 563–579.

Samter, M. Intolerance to aspirin. *Hospital Practice*, 1973, **12**, 85–90.

Shaffer, D. Psychiatric aspects of brain injury in childhood: A review. *Developmental Medicine and Child Neurology*, 1973, **15**, 211–220.

Silbergeld, E. K., & Goldberg, A. M. A lead-induced behavioral disorder. *Life Sciences*, 1973, **13**, 1275–1283.

Silbergeld, E. K., & Goldberg, A. M. Lead-induced behavioral dysfunction: An animal model of hyperactivity. *Experimental Neurology*, 1974, **42**, 146–157.

Silver, L. B. Familial patterns in children with neurologically-based learning disabilities. *Journal of Learning Disabilities*, 1971, **4**, 349–358.

Sobotka, T. J. Estimates of average, 90th percentile and maximum daily intakes of FD & C artificial food colors in one day's diets among two age groups of children. Food and Drug Administration memorandum, July 1976.

Speer, F. The allergic tension–fatigue syndrome in children. *International Archives of Allergy*, 1958, **12**, 207–214.

Speer, F. *The allergic child*. New York: Harper & Row, 1963.

Spring, C., Vermeersch, J., Blunden, D., & Sterling, H. *Case studies of effects of artificial food colors on hyperactivity*. Unpublished manuscript, University of California, Davis, Calif.

Stewart, M. A., & Morrison, J. R. Affective disorder among the relatives of hyperactive children. *Journal of Child Psychology and Psychiatry*, 1973, **14**, 209–212.

Swanson, J. M., & Kinsbourne, M. Stimulant related state dependent learning in hyperactive children. *Science*, 1976, **192**, 1354-1357.

Swanson, J. M., & Kinsbourne, M. *Artificial food colors impair the learning of hyperactive children*. Report to the Nutrition Foundation, 482 Fifth Avenue, New York, N.Y., 1979.

Swanson, J. M., & Logan, W. *The effect of food dyes on neurotransmitters*. Report to the Nutrition Foundation, 482 Fifth Ave., New York, N.Y., 1979.

Swanson, J. M., Kinsbourne, M., Roberts, W., & Zucker, K. Time response analysis of the effect of stimulant medication on the learning ability of children referred for hyperactivity. *Pediatrics*, 1978, **61**, 21–34.

Sykes, D. H., Douglas, V. I., & Morgenstern, G. Sustained attention in hyperactive children. *Journal of Child Psychology and Psychiatry*, 1973, **14**, 213–220.

Tryphonas, H., & Trites, R. Food allergy in children with hyperactivity, learning disabilities and/or minimal brain dysfunction. *Annals of Allergy*, 1979, **42**, 22–27.

Ulett, G. A., & Perry, S. G. Cytotoxic testing and leukocyte increase as an index to food sensitivity. *Annals of Allergy*, 1974, **33**, 23–32.

Vaughan, W. T. Food allergens. III: The leucopenic index, preliminary report. *Journal of Allergy*, 1934, **6**, 601–605. (a)

Vaughan, W. T. Further studies on the leucopenic index in food allergy. *Journal of Allergy*, 1934, **6**, 78–85. (b)

Vaughan, W. T. Food idiosyncrasy as a factor in the digestive leucocyte response. *Journal of Allergy and Clinical Immunology*, 1935, **5**, 421–430.

Vaughan, W. T. The leucopenic index as a diagnostic method in the study of food allergy. *Journal of Laboratory and Clinical Medicine*, 1936, **1**, 1278–1288.

The Washington Post, October 29, 1973.

Weiss, B., Margen, S., Williams, J. H., Abrams, B., Caan, B., Citron, L. J., McKibben, J.,

Ogar, D., & Schultz, S. *Final report on Phase 2* FDA Contract No. 223–76–2040, December 1978.

Weiss, G. The natural history of hyperactivity in childhood and treatment with stimulant medication at different ages. *International Journal of Mental Health*, 1975, **4**, 213–226.

Weiss, G., Minde, K., Werry, J. S., Douglas, V. I., & Nemeth, E. Studies on the hyperactive child VIII: Five year follow-up. *Archives of General Psychiatry*, 1971, **24**, 409–414.

Werry, J. S. Food additives and hyperactivity. *Medical Journal of Australia*, 1976, **2**, 281–282.

Werry, J. S., & Sprague, R. L. Methlyphenidate in children. Effect of dosage. *Australian and New Zealand Journal of Psychiatry*, 1974, **8**, 9–19.

Werry, J. S., Sprague, R. L., & Cohen, M. N. Conners' teacher rating scale for use in drug studies with children: An empirical study. *Journal of Abnormal Child Psychology*, 1975, **3**, 217–229.

Williams, J. I., Cram, D. M., Tausig, F. T., & Webster, E. Relative effects of drugs and diet on hyperactive behaviors: An experimental study. *Pediatrics*, 1978, **61**, 811–817.

WOR Radio–TV reports. *The McCann Show*, WOR Station, New York, June 18, 1974, 11:15 A.M.

INDEX